CONCEPTS OF ADVANCED ZERO WASTE TOOLS

CONCEPTS OF ADVANCED ZERO WASTE TOOLS

Present and Emerging Waste Management Practices

Edited by

CHAUDHERY MUSTANSAR HUSSAIN, Ph.D.
Department of Chemistry & Environmental Science
New Jersey Institute of Technology
Newark, NJ, United States

ELSEVIER

Elsevier
Radarweg 29, PO Box 211, 1000 AE Amsterdam, Netherlands
The Boulevard, Langford Lane, Kidlington, Oxford OX5 1GB, United Kingdom
50 Hampshire Street, 5th Floor, Cambridge, MA 02139, United States

Library of Congress Cataloging-in-Publication Data
A catalog record for this book is available from the Library of Congress

British Library Cataloguing-in-Publication Data
A catalogue record for this book is available from the British Library

ISBN: 978-0-12-822183-9

Publisher: Candice Janco
Acquisitions Editor: Marisa LaFleur
Editorial Project Manager: Leticia M Lima
Production Project Manager: Debasish Ghosh
Cover Designer: Miles Hitchen

Typeset by SPi Global, India

Dedication

I would like to dedicate this book to
My beloved GOD
"Meray Pyarey Allah"

Contents

Contributors *xi*
About the editor *xiii*
Series preface to first edition *xv*
Preface *xvii*

1. The realm of zero waste technology: The evolution **1**

Sunpreet Singh, Seeram Ramakrishna, and Chaudhery Mustansar Hussain

1. The waste 1
2. History of ZWT 4
3. Present age of ZW 10
4. Conclusions 17

References 17

2. Zero waste certification **23**

Kailash Shweta Pal, S. Subhashini, and Kantha Deivi Arunachalam

1. Introduction 24
2. What is zero waste? 26
3. What is the zero waste strategy? 27
4. What is zero waste certification? 27
5. Why ZWC is to be done? 28
6. What are the benefits of ZWC? 30
7. What are the standard for certification? 31
8. Who certifies zero waste? 33
9. Zero waste certificate initiatives around the world 37
10. Conclusion 41

References 41

3. Zero waste manufacturing **45**

Sunpreet Singh and Chaudhery Mustansar Hussain

1. Introduction 45
2. Prospective of ZWM 48
3. Innovative industrial implications: Studies with ZWM view-points 54
4. Conclusions 61

References 62
Further reading 67

4. Challenges, issues, and problems with zero-waste tools **69**

K.S. Vignesh, Suriyaprakash Rajadesingu, and Kantha Deivi Arunachalam

1. Introduction 69
2. The concept of "zero waste" systems 71
3. The ZW development 75
4. The key aspects of developing a ZW strategy 79
5. Conclusion 87

References 88

5. Application of poultry industry waste in producing value-added products—A review **91**

Sathish Kumar Karuppannan, Mohammed Junaid Hussain Dowlath, G.I. Darul Raiyaan, Suriyaprakash Rajadesingu, and Kantha Deivi Arunachalam

1. Introduction 91
2. Poultry processing steps and wastes generated 93
3. Zero-waste approach in handling poultry waste 95
4. Conclusion 114

Important websites 114

Reference 115

6. Zero waste hierarchy for sustainable development **123**

Sunpreet Singh and Chaudhery Mustansar Hussain

1. Zero Waste concept: From literature's viewpoint 123
2. ZW hierarchies: Sustainable behavior 127
3. Case studies: Success stories of ZW 135
4. Conclusions 138

References 139

Further reading 142

7. Application of advanced technologies in managing wastes produced by leather industries—An approach toward zero waste technology **143**

Mohammed Junaid Hussain Dowlath, Sathish Kumar Karuppannan, Pratheeka Rajan, S.B. Mohamed Khalith, Suriyaprakash Rajadesingu, and Kantha Deivi Arunachalam

1. Introduction 144
2. Processing steps involved in leather production 145

3. Overview of leather industry waste 147
4. Zero-waste approach 152
5. Technological solutions for the challenges in leather industry 153
6. Conclusion 169
Important websites 171
References 171

8. Modern society and zero waste tools 181

Suriyaprakash Rajadesingu, V. Deepankara, Mohammed Junaid Hussain Dowlath, Sathish Kumar Karuppannan, and Kantha Deivi Arunachalam

1. Introduction 181
2. Barriers toward achieving zero waste 182
3. The notion of the "zero waste city" 183
4. The concept of zero waste and the zero-waste city 184
5. Key steps toward zero waste cities: Zero waste management 185
6. Waste treatment and disposal 185
7. Regulatory strategies 186
8. Governance and infrastructure 186
9. Market creation 186
10. Global waste issues and cities: Why zero waste? 187
11. Current and future cities: A crucible of issues—A milieu for innovation and opportunity 192
12. Zero waste: Formation, convergence, circularity, and critique 194
13. Who-how-what-why: Zero Waste? 196
14. Conclusions: Zero waste and the design of future cities 200
References 204
Further reading 212

9. Toward "Zero Liquid Discharge" industrial facilities: Reducing the impact on freshwater resources by reusing industrial and urban wastewaters 215

Antonio Santos Sánchez

1. An introduction to zero liquid discharge 215
2. Reuse of municipal wastewater 217
3. A successful case study: Aquapolo reclamation plant 220
4. Outcomes of the Aquapolo project 232

5. How the Aquapolo project helped fostering water efficiency and internal reuse projects among the participating companies 235
6. Conclusions 242
References 243

10. Conclusions 247
Sunpreet Singh and Chaudhery Mustansar Hussain

Index 249

Contributors

Kantha Deivi Arunachalam
Department of Biotechnology, School of Bioengineering; Center for Environmental Nuclear Research, Directorate of Research, SRM Institute of Science and Technology, Kattankulathur, Tamil Nadu, India

V. Deepankara
Center for Environmental Nuclear Research, Directorate of Research, SRM Institute of Science and Technology, Kattankulathur, Tamil Nadu, India

Mohammed Junaid Hussain Dowlath
Center for Environmental Nuclear Research, Directorate of Research, SRM Institute of Science and Technology, Kattankulathur, Tamil Nadu, India

Chaudhery Mustansar Hussain
Department of Chemistry & Environmental Science, New Jersey Institute of Technology, Newark, NJ, United States

Sathish Kumar Karuppannan
Center for Environmental Nuclear Research, Directorate of Research, SRM Institute of Science and Technology, Kattankulathur, Tamil Nadu, India

S.B. Mohamed Khalith
Center for Environmental Nuclear Research, Directorate of Research, SRM Institute of Science and Technology, Kattankulathur, Tamil Nadu, India

G.I. Darul Raiyaan
Center for Environmental Nuclear Research, Directorate of Research, SRM Institute of Science and Technology, Kattankulathur, Tamil Nadu, India

Suriyaprakash Rajadesingu
Center for Environmental Nuclear Research, Directorate of Research, SRM Institute of Science and Technology, Kattankulathur, Tamil Nadu, India

Pratheeka Rajan
PG and Research Department of Biotechnology, Women's Christian College, Chennai, India

Seeram Ramakrishna
Mechanical Engineering, National University of Singapore, Singapore, Singapore

Antonio Santos Sánchez
Federal University of Ouro Preto, Ouro Preto, Brazil

Kailash Shweta Pal
Department of Biotechnology, School of Bioengineering, SRM Institute of Science and Technology, Kattankulathur, Tamil Nadu, India

Sunpreet Singh
Mechanical Engineering, National University of Singapore, Singapore, Singapore

S. Subhashini
Department of Biotechnology, School of Bioengineering, SRM Institute of Science and Technology, Kattankulathur, Tamil Nadu, India

K.S. Vignesh
School of Public Health, SRM Institute of Science and Technology, Chennai, India

About the editor

Chaudhery Mustansar Hussain, PhD, is an Adjunct Professor and Director of Labs in the Department of Chemistry & Environmental Sciences at the New Jersey Institute of Technology (NJIT), Newark, New Jersey, United States. His research is focused on the applications of Nanotechnology & Advanced technologies & Materials, Analytical Chemistry, Environmental Management, and Various Industries. Dr. Hussain is the author of numerous papers in peer-reviewed journals. He is a prolific author and is the editor of several scientific monographs. Handbooks in his research areas are published with Elsevier, Royal Society of Chemistry, John Wiley & Sons, CRC, Springer, and many others.

Series preface to first edition

Recently the concept of zero waste has become a topic of considerable importance. It is significant not only to environmental managers but also to environmental engineers and scientist, chemists, toxicologist, biotechnologist, pharmacists, forensic scientist, and environmental regulators who need to reduce/eliminate their waste and has evolved as a true discipline throughout the world. Zero waste is basically a philosophy that encourages the redesign of resource life cycles so that all products are reused. The goal is that no trash should be sent to landfills, incinerators, or the ocean. The process recommended is one similar to the way that resources are reused in nature. It is perhaps the most powerful and versatile concept available to date for an environmentalist.

Every day the new developments in environmental sciences and engineering, spread of industrialization, and growing global population increase the waste production. The increase in waste production increases the need for more areas to dispose and new technology to design which is limiting our resources from the environment. To relieve the pressures placed on the finite resources available, it has become more important to prevent waste. Zero waste promotes not only reuse and recycling but also, more importantly, it promotes prevention and product designs that consider the entire product life cycle. Zero waste designs strive for reduced materials use, use of recycled materials, use of more benign materials, longer product lives, reparability, and ease of disassembly at end of life. In general, zero waste strongly advocates the sustainability by protecting the environment, reducing costs, and usage of wastes back into the industrial cycle. However, until today, the advance comprehensive understanding and real world concept zero waste tools are still a challenge. This series addresses these challenges of implementation of zero waste tools at both real and conceptual model scales.

Overall, concept of zero waste is a goal, a process, a way of thinking that strongly changes our approach to resources and production. It is not about recycling and diversion from landfills but about restructuring production and distribution systems to prevent waste from being manufactured in the first place. The materials that are still required in these redesigned resource-efficient systems will be reused many times as the products that incorporate them are reused. This series summarize present and emerging concept zero waste tools, at both experimental and theoretical models scales.

Moreover, economical, toxicological, and regulatory issues will be presented in detail. In the end, the research trends and prospective in the future will be briefly debated.

Chaudhery Mustansar Hussain
Series Editor

Series volumes

Volume 1: Concept of Advanced Zero Waste Tools
Volume 2: Source Reduction & Waste Minimization
Volume 3: Waste-to-Energy: Approaches Towards Zero Waste
Volume 4: Emerging Trends to Approach Zero Waste
Volume 5: Bio-Based Materials & Technologies to Approach Zero Waste

Preface

Zero waste tools (ZWT) is a conceptual approach to acquire a shift from the common traditional waste management model to integrated systems in which everything has its value and usage in environment. It advocates an urbanized transformation that can minimize its impact on the natural resources. Zero waste can be categorized into subsystems such as zero waste in administration and manufacturing, zero waste of resources, zero emissions, zero waste in product life, and zero use of toxics. Although it has been long since the emphasis on the ZWT has started, however, only a few implication have been brought in practice at the ground level, especially within micro- and small-scale industries. Zero waste manufacturing is a theoretical word that is in demand among all the top-notch manufacturing domains across the world. The manufacturing wastes are abundant and versatile in term of their category involving plastics, metals, ceramics, and others. Hierarchy of zero waste concept through a comprehensive emphasis on the associated sustainability requisites. The major focus has been made on presenting hierarchy, as a concept, to promote waste avoidance ahead of recycling and disposal. In particular the 6R's, such as reconsider, reuse, reduce, recycle, recover, and retain, have been well explained with respect to their physical, social, and economic relationships through the medium of established theories and practices available in the literature. Overall the ZWT is both an inspiration and a resource terminology to help industries and societies to collaborate on the adoption of sustainable practices, which can reduce the waste and work for the circular economy using in-process material wastages. In this volume, we summarized modern developments in various concepts of advanced ZWT.

ZWT is a set of principles focused on waste prevention that encourages the redesign of resource life cycles so that all products are reused, being an ultimate goal that no trash should be sent to landfills, incinerators, or the ocean. This book provides invaluable insights of the broad horizons of several concepts the ZWT practiced so far in the various commercial sectors.

Overall, this book is designed to be a reference guidebook for experts, researchers, and scientists who are searching for new and modern concepts of advanced ZWT. The editor and authors are well-known researchers, scientists, and professionals from academia and industry. On behalf of Elsevier, we are pleased with all the authors for their outstanding and passionate hard work extended toward the completion of this book. We extend our extraordinary acknowledgements to Marisa LaFleur and Peter Llewellyn (acquisition editors) and Letícia Lima (editorial project manager) at Elsevier, for their dedicated support and help during this project. In the end, we thank Elsevier for publishing this book.

Chaudhery Mustansar Hussain
Editor

CHAPTER ONE

The realm of zero waste technology: The evolution

Sunpreet Singh[a], Seeram Ramakrishna[a], and Chaudhery Mustansar Hussain[b]
[a]Mechanical Engineering, National University of Singapore, Singapore, Singapore
[b]Department of Chemistry & Environmental Science, New Jersey Institute of Technology, Newark, NJ, United States

Contents

1 The waste 1
2 History of ZWT 4
  2.1 Evolution of ZW and ZWT 6
3 Present age of ZW 10
  3.1 Sustainability 11
  3.2 Innovative technologies 13
4 Conclusions 17
References 17

1 The waste

The United Nations (UN) defined waste as "the materials which not essential products and has no further use and ready to dump-off." The generation of waste is inevitable as it can be generated during raw materials' extraction, processing of raw materials into intermediate/final products, consumption of final products, and other human-based activities (United States Environmental Protection Agency, 2011). Further, there exist various other types of wastes that are inextricably linked to their solid state, and the UN stated that "it is a primary aim of wastewater treatment is removing solids from the wastewater" (United Nations Environment Programme, 2011). The scientific knowledge of the literature contains many references to the overabundance and dangers of waste. According to Barnes et al. (Barnes et al., 2009), the overall global rate of municipal solid waste (MSW) generation is estimated to be 1.2 kg/person/day in 2010, and this is predicted to increase to 1.4 kg/person/day by 2025.

Concepts of Advanced Zero Waste Tools
https://doi.org/10.1016/B978-0-12-822183-9.00001-5

Considering the increase in the population, the total generation of MSW is expected to increase from 1.3 billion tons/year in 2010 to 2.2 billion tons/year in 2025. With regard to MSW, it is equal to a global increase of 40 tons per second in 2010 and may increase to 70 tons per second by 2025. According to Danilov-Danil'yan et al. (2009), the observed environmental pollution because of solid waste can turn into the main threat due to modern civilization. In addition, the risk posed by the abundance of synthesized chemicals in the waste stream has allowed no organisms to evolve in nature to break them down and render them harmless (Meadows et al., 2005). In the future millennia, archeologists will face difficulty in identifying human civilization as an enormous amount of nondegraded garbage will have been buried. Further, plastics will likely be our most visible legacy as these materials have longevity estimated for hundreds of thousands of years (Barnes et al., 2009; Weisman, 2007).

Rios et al. (2007) conducted a study of plastic debris retrieved from various locations in the Pacific Ocean that confirmed that this material is a trap for persistent organic pollutants. Further, Meadows et al. (2005)) noted that over 65,000 industrial chemicals are now in regular commercial use that possesses limited or no toxicology data available for them as a result of limited testing. In the category of hazardous wastes, none is critically problematic, for example, the by-products of nuclear energy and weapons development. However, this problem is neither political nor scientific that can mitigate. Further, numerous security risks associated with nuclear or weapon materials have several environmental challenges. A similar category of wastes, for example, e-waste refers to electronic technologies that have been produced from computers, televisions, and cell phones are other common sources of environmental hazards. In (Carroll, 2008), it has been cited that in the United States, about 70% of discarded computers and 80% of discarded TVs end up in landfills. The National Safety Council estimated that the United States itself produced around 250 million discarded computers during 2004–2009 (Royte, 2005). However, the ongoing technological advances guaranteed routine uselessness, and the peaks of the e-waste mountain are continuously growing. Indeed, the growth of e-waste is spilling over from the developed world, and the less developed countries demand international efforts and treaties to prevent it.

Puckett et al. (2002) reported that a million pounds of e-waste from obsolete computers and TVs are being produced in the United States and an estimated 80% of this has been collected for recycling and exported to other countries. The consequences of the e-waste trade in Asia, China,

India, and Pakistan are extremely polluting and are likely to cause serious damage to human health. The e-waste is, therefore, an example of how the dangers of waste flow along a general gradient from wealthy to poor. Similarly, burying the rubbish in a hole in the ground, better known as landfilling, is the most common modern means of dealing with it; however, the dangers posed by this are well documented.

Girling (2005) presented an example of a British study where they found that children born within 2 km of a landfill site were statistically more likely to suffer from abnormalities. Watson (2009) reported that waste management (WM) processes are a major source of greenhouse gas emissions through the decomposition of organic matter in landfill sites, which produces the greenhouse gas methane. In addition, the landfilling sites have their proliferation steadily subtracted from the earth's finite land base. Since 2001, the NY City has been shipping the rubbish out of state to NJ City, Pennsylvania, and Virginia, more than 500 km away. Similarly, Toronto's local landfill filled up in 2002, and it is now exporting its waste across the border into Michigan (Brown, 2008).

Incineration presents an alternative to landfilling, and as per the notion "burning waste will make it to go away" is a misconception. Danilov-Danil'yan et al. (2009) pointed out that the incineration decreases about 90% of solid waste volume and transforms it into a gas. It has been found that a single ton of solid waste creates about 30 kg of air-borne ashes and 6000 m^3 of fume gases containing sulfur dioxide, nitrogen and carbon dioxide, hydrocarbons, heavy metals, and dioxins, respectively. Moreover, at 90% reduction in solid waste volume, there is still the 10% residual volume of toxic incinerator ash that requires landfilling—so incineration is ultimately unable to eliminate the dependence on finding new landfill space. Several other examples of waste management point out the fact how waste can be dealt with. The four steps that convert virgin materials into waste include extraction and manufacture, distribution, consumption, and waste. This transformation takes place rapidly in a developed society.

The waste is a problem on a significant and global scale, and if left unchecked, the risks to humans and overall ecosystem can be severe. Levin et al. used the term *super wicked* to describe wicked problems of the severest degree, as an issue of global climate change. They describe: shortage of time, problem originator should also provide solution, weak and nonexistent address by government authorities, and irrational discounting system to push responses to future. Levin et al. (2012) argued that the combination of features comprising a wicked problem has created a tragedy because of our governance institutions and policies. Since the time is demanding the potential solutions to the problems of waste, thus, Section 2 explores the historic milestones set by zero waste technology (ZWT) movement.

2 History of ZWT

The origin of the ZWT terminology is difficult to locate as it appears that its first institutional use was in the late 20th century when an American chemist Paul Palmer founded ZW systems, a commercial enterprise specialized in the reuse of chemical by-products that were otherwise destined for disposal (Palmer, 2007). During the 1990s, the term had been widely adopted by the grassroots activists in English-speaking countries around the world, with analogous terms such as Zéro Déchet (French) and Null-Abfall (German) used in other languages in different countries. Literally, ZW means a complete, 100% elimination/absence of waste. Further, the literature revealed that ZW is often not interpreted literally. To be able to assess ZW initiatives around the world, it is necessary to understand the different meanings imparted to this term. Kozlowski Russell (2009) conducted a review of prominent ZW definitions and observed that the terms serve many functions at once, for example:

- waste reduction goal,
- visionary statement,
- resource management, and
- solution to pollution and global climate change.

The ZW is, therefore, a multifaceted topic to bring in a host of other terminology and intended meanings as well. It is of the utmost importance to understand how all of this language is interrelated (Glavič and Lukman, 2007). It has been observed that terms such as ZW, commonly used to describe societal goals, are not ideal. Murray (Murray, 2002) suggested that ZW is a contradiction as:

> *Waste is defined as matter in the wrong place and eliminating waste can take with it the possibility of matter being in the right place.*

A study of ZW revealed that an important part of understanding this term involved what are the irrelevant terms. For example, Havel et al. (2006) asserted that ZW means the reduction of the production of all types of waste to zero, which in reality is not possible in a society oriented toward consumption. Further, the term defining the elimination of the present methods of waste disposal, such as landfilling or incineration, is also impossible.

Reportedly, ZW is not a technology but a strategy that begins with better industrial design and ends with source separation of discarded products. As per the Central Vermont Solid Waste Management District (2009), ZW is not about eliminating discards but to strive to capture the resources in the rubbish products in a way that they can be reused and recycled instead of landfill or incinerated. According to Snow (2011), ZW has been compared with the older manufacturing sector goals, including zero emissions, zero accidents, and zero defects. It has been pointed out by the researchers that all of these were adopted as "impossible" targets initially, but proved their worth by dramatically changing industry and society. The assertion has been supported by Edgerly and Borrelli (2007), who note that more modest levels of waste reduction such as 50%–60% diversion by some municipalities and regions have been helped by having ZW goals in place.

Admittedly, the definitions of ZW vary from those that are aspirational only and may resemble WM practices. For example, in Karani and Jewasikiewitz (2007) the author highlighted that the most understood definition of ZW is the minimization of waste generation, the reuse and recycling of waste, and the diversion of waste away from landfill or incineration. ZW acknowledges more than minimization, reuse, recycling, and diversion. ZW is a full-system view to focus on the recapture of the resources from the waste stream, decreasing consumption (Dinshaw et al., 2006). ZW is also defined as a process where the materials destined for landfills or incinerators are returned upstream to be recycled as feedstock for new products or services (Doppelt and Dowling-Wu, 1999), or else are naturally decomposed so they can be reintegrated into nature without environmental impacts. Lehmann (2011a) outlined that ZW challenges the most common assumptions:

> *Waste is unavoidable and has no value.*

A Grassroots Recycling Network (2009) described ZW:

> *A concept to redesigns the current, one-way industrial system into a circular system modelled on Nature's successful strategies.*

Connett and Sheehan (Connett and Sheehan, 2001) pointed out:

> *The need to reconfigure our one-way industrial system into a circular, closed-loop system.*

However, Edgerly and Borrelli (2007) asserted:

> *ZW offers a circular resource management system in which discarded materials are looped back into the economy.*

It has been seen, till now, that the meaning of the ZW terminology has considerable variations in terms of what exact meaning is intended from case to case. Overall, two themes stand out:

- ZW represents a paradigm shift and is beyond merely finding better variations to the same old WM strategies.
- ZW considers waste as a resource of the circular system instead of an externality that is the end product of a one-way, linear system.

ZW is the combination of various philosophical goals to reduce, or even eliminate, the waste. However, in different ways, ZW can be articulated, and a commonly found view in the literature is that at least a partial part of the necessary change is required to successfully address the global waste problem. Countries, such as China, India, and Indonesia, are attempting to catch up with the western consumption and suggesting that the stresses on finite resources threaten to become worse. The ZW International Alliance (Zero Waste International Alliance, 2009) suggested:

> *ZW is a goal that is ethical, economical, efficient and visionary, to guide people in changing their lifestyles and practices to emulate sustainable natural cycles, where all discarded materials are designed to become resources for others to use.*

Section 2.1 takes a close insight at how ZW initiatives around the world have taken up the waste and are discussed in detail.

2.1 Evolution of ZW and ZWT

The environmental issues evolved due to the rapid economic development around the world include depletion of natural resources, large amounts of industrial wastes, pollution with radioactive and toxic elements, lowering the fertility of the land, reduction of land, limited production efficiency, and many others. (Tyulenev et al., 2017). The negative impact of the human activities on environmental conditions relies on surveying the large-scale introduction of the ZWT technologies to help solving problems of air pollution, hydrosphere, lithosphere pollution, and reduce the amount of garbage (Tyulenev et al., 2018; Koryakov and Kulikov, 2018). The positive environmental effect of ZWT is immense (Ohrn-Mcdaniel, 2014).

The new creative approaches, in the area of fashion and patternmaking, are more difficult to create a marketable design that is easy to reproduce. While designing garments, it is most essential to reproduce the garment in a balanced timeframe to fit within an appropriate cost. The ZW approach and applying it to marketable design for the bridge market created a box for the designer to work within. To contradict the traditional shapes of ZW technology, the adopted technology must be successfully used in production. Some authors described ZWT as a model to meet the following credentials (Upadhyaya, 2013):

- Lifecycle design: This includes the design and development of commercial good or service in a way that it can meet the entire lifecycle of the products. In a simple definition, the development of a product so that it can be used for the entire lifecycle.
- Degree of demonstration: It emphasizes the fact that forbidding plastic baggage is not an option. Rather, the aim should be on the alternate, not on ostracize.
- Designing for upstream: The correct style starts once a product is anticipated, even as prices and markets to end with degraded or mixed garbage.
- Capturing the highest functioning: the materials should not focus on the repair and use of each product for the highest available price.
- Highest manifestation and reusability: The buildings, cooling systems, processing plants, conveyor systems, and others should be preserved as complete systems and reinstalled when needed.

The circular economy (CE) concept, in the context of ZW, is an interwoven concept that predates ZW, for example, industrial symbiosis (Chertow and Ehrenfeld, 2012). The eco-friendly city is a CE-related concept that is evolving in Japan, Singapore, and other parts of the world (Dong et al., 2016). In industries, the CE concept is associated with a broad range of subjects such as thermodynamics and ecological economics (Winans et al., 2017). In systems theory and according to thermodynamics, the application of CE concepts influences the production and consumption models in a way that enables a "*de-growth phase*" of the economic system (Jiao and Boons, 2017). Apart from the growth of the CE, a business paradigm toward sustainability, the concept has been explored at different levels but not so much debated academically within business and sustainability literature (Murray et al., 2017). Further, the term sustainability cannot be fully associated with the CE because the social component of sustainability is practically dismissed (Schneider, 2015). The ZW concept in the past looked at eliminating the unneeded use of materials and the optimization of resources; following rethink, reduce, reuse, recycle, recover, residuals management, and leaving unacceptable (Hanna, 2015).

WM, in the earlier decades, has been changed through environmental issues and concepts of waste at certain periods, illustrated by the keywords including cleaning, pollution protection, producer responsibility, resources, and renewable energy in Korea (Yang et al., 2015). Since the legislation began in the 1960s, different laws and acts have been enacted in WM based on such issues in history. It has been seen that the highest level of changes occurred in the waste depends on categorizing the waste and its application. Changes that occurred in the classification of WM have produced several efficient results such as an increase in the recycling rate, ease of handling waste stream, and decreasing treatment costs. Until the WM Law came into effect in December 1986, the Filth and Cleaning Law (1973) and the Environmental Protection Law (1963) have been regulated in Korea (Pariatamby and Tanaka, 2014). On the universal level, the various milestones set have been made as discussed in (Connett, 2013). The scope of the ZW study conducted by Zaman (2017) highlighted that all ZW life phase from the extraction of resources to the final disposal of waste. Researchers conducted a significant number of studies on regulatory policy (19%); however, the greatest number of studies were conducted on WM (22%) followed by ZW extraction processes (32 studies) and ZW treatment (31 studies). There should be a significant change to make on the refurbishment of human's life style to reduce consumption (Hogland et al., 2014). Following this concept, environmentally friendly ways of living using alternative sources of energy and eating organically grown food have already been started among the literate community. For accepting that the global problem of waste is extremely challenging, one should recognize the cost to society for addressing the most acute issues (Wilson et al., 2015). Relatively, the generation of the progressive sound economic investment, the socioeconomic and environmental opportunity of addressing waste issues should be considered as strengths. However, steps in producing the ZW strategy start with the preliminary assessments (refer Fig. 1) of existing WM systems. The preliminary assessment and evaluation is important to measure existing WM performance. The waste characterization and prominent issues in achieving ZW goals need to be identified at this preevaluation stage (Kerdlap et al., 2019).

The audit and reduction of the technologies, in a smart way, are feasible to implement on the software side for waste data management and benchmarking. Indeed, the physical process of waste collection is the technical barrier that needs to be overcome by adopting digital means. Despite the top economies of the world, such as Japan, the United States, and the United Kingdom, developing part of the world also sorts wastes onsite through specialized bins. Further, smart waste bins can measure the overall volume of waste generated, but they cannot measure the volumes of waste

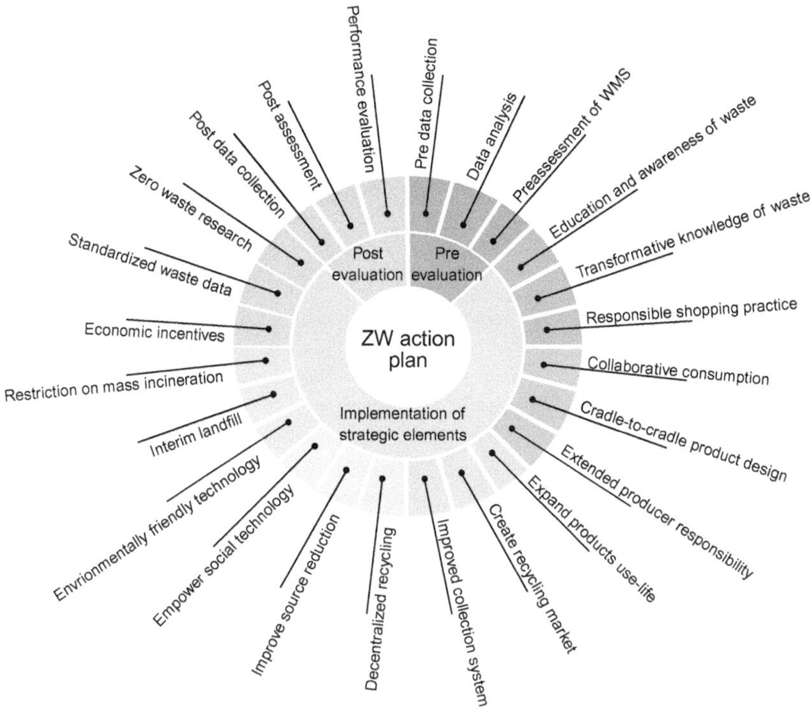

Fig. 1 Steps in developing a ZW strategy (Zaman, 2017).

by material type. The dense urban centers, like Singapore, require hardware technologies that can efficiently measure the volume of waste generated and estimate its material composition (Hogland et al., 2014). Any legislation or legal policy for the ZW should be introduced by adopting the overarching guidelines implemented in four phases (refer Fig. 2). The prime motive of such resulting strategies, despite overcoming the waste, should also reduce the material and energy consumption in all nations (Lehmann, 2010). For achieving the ZW, the main difficulty requires continued and combined efforts by industry, government bodies, university researchers, and the people and organizations in our community. Further, the governments should formulate the effectiveness of the policies to reduce the environmental impacts of consumption and production, addressing issues such as household consumption, public procurement, corporate behavior, and technological innovation. Zaman (2015) outlined some recommendations to follow.

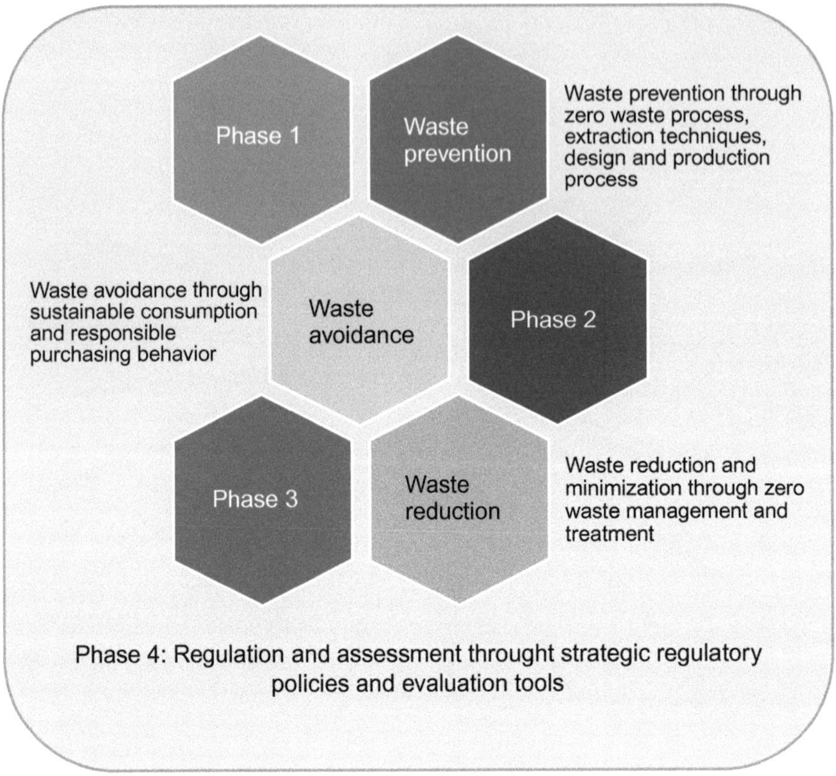

Fig. 2 Phases in implementing ZW strategy (Zaman, 2015).

It has been understood that waste control can be started at individual levels; however, various business groups and concerned government authorities can help the individuals to make more environmentally conscious methodologies. There are many success stories around the world including Germany, Austria, Spain, Italy, the United Kingdom, Singapore, Korea, Japan, and the United States. Similarly, the developing world is still worrying about financial troubles.

3 Present age of ZW

The philosophical concept of ZW and ZWT has evolved into several technical methodologies providing realistic checks to evaluate whether the implemented systems are providing qualitative and/or quantitative outcomes for evaluations. This section aimed to review the guiding principles, particularly in the context of:

- sustainability goals: including the complex relationships between impacts systems and
- new innovative technologies.

3.1 Sustainability

The term "Sustainability" is defined as the goal of sustainable development including economic and social development that protect and enhance the natural environment and social equity (Gertsakis and Lewis, 2003). This term entered the public debate after the World Commission on Environment and Development published their landmark report, Our Common Future, in 1987. It was defined by the report that the development meets the needs of today's requirements without compromising the ability of future generations (WCED (World Commission on Environment and Development), 1987). Sustainability concept recommends that the industries and their operations should be encouraged to become more efficient in terms of utilization of the resources, production of less pollution and process waste, and the use of nonrenewable resources to minimize irreversible adverse impacts on human health and the environment. Further, the sustainable society should be the one that persists overwhelming generations, flexible, and wise to undermine either its physical or its social systems of support. The sustainability should focus on the use of renewable resources, sustainable alternatives of renewable substitutes, and generation of the wastes with only assimilative capacity.

For sustainable development, it is of utmost importance to establish a waste hierarchy for WM, at a global level, that can influence the technical trajectories of WM regimes. Fig. 3 shows the hierarchy of WM. The development of new technological approaches can serve as a launching pad for specific technological trajectories and the basis for a more sustainable organization of economic activities (Buclet and Godard, 2013). The suggested hierarchy should stem from; first, in the short/medium term, this approach can reduce the benefits of management options that must be better to adapt specific waste streams (Lazarevic et al., 2010). However, the context of the waste control hierarchy has been changed with the geographical locations, as represented in (Williams, 2015). Despite the increasing reuse and recycling efforts, the majority of waste produced is still placed into storage facilities. The "*R-word*" in scientific literature has proliferated and delivered the

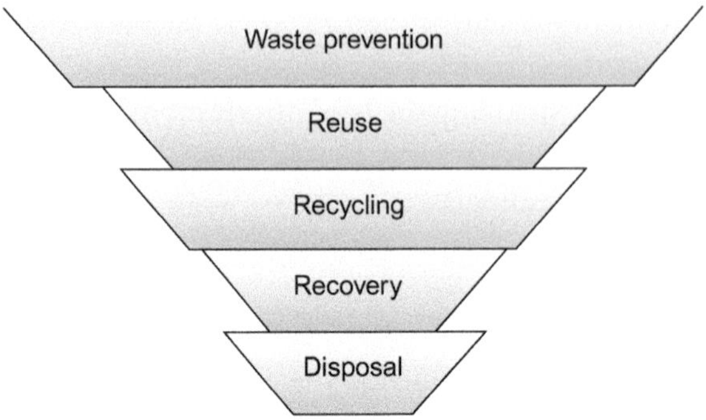

Fig. 3 Waste management hierarchy.

key aspects of WM and resource recovery. Further, these have been used freely along with reprocessing, recycling, resource recovery, and reuse, as well as rehabilitation, reclamation, restoration, recuperation, reconstruction, and remediation (Lottermoser, 2011).

It is critical to be aware and develop better measures and goals for moving toward ZW and CE to emphasize on source reduction, reduced consumption, product and service redesign, and reuse. Veleva et al. (2017) described how to measure something that is nonexisting. For identifying the reuse factor, it is a bold task yet very fewer enterprises are tracking such records. In most of the cases, an outbound party hired by the enterprise collects the used products, and no information is relayed back. It has been concluded by Cole et al. (2014) that ZW is difficult to achieve without enforcing clear management policies, and the zero waste strategies should include social and environmental aims with WM performance targets and the ability to monitor progress. To achieve set milestones, there is a need to establish a link between all stakeholders to produce a holistic approach to WM.

Further, this requires additional efforts, innovation, creative and effective policies, partnership working, and support from National Governments to move in the direction of ZW. Similarly, the ZW model in emerging countries should be developed for post-consumer waste in urban areas with differing levels of service. Indeed, waste minimization and recycling have to be identified as necessary steps for the success of any strategic model. These should also focus on the waste raised after the minimization and recycling of such wastes using a suitable model (Matete and Trois, 2008). In addition, there are lots of suggestions given by the authors mentioned in Couth and Trois (2012), Zaman (2014), Lehmann (2011b), Carrico and Kim (2014), and García-Ten et al. (2016). In recent years, environmentalists of the developed world are criticizing the increase of manufacturing and household wastes and advocating the institutionalization of ZW as a societal goal to spur WM agencies to eliminate the need for landfills. Further, the integrated WM should recognize landfilling as the last choice available in the hierarchy of WM (Navia and Ross, 2009).

Further, several solutions have been suggested to improve the use of the hierarchy with regard to achieve dematerialization (Van Ewijk and Stegemann, 2016). The hierarchy can be comprehensively specified on the open-loop and closed-loop recycling, along with for different combinations of materials and treatments, including bans on landfill of particular materials and products. The CE concept as background, the waste hierarchy index (WHI), can be upgraded to allow the calculation of the level of implementation considering recycling and incineration, and the finally modified system can contribute to CE. The streamlined WHI allows its application at the moment, being valid to be applied with existing data from European countries reporting to Eurostat (Pires and Martinho, 2019). Cherubini et al. (2008) investigated scenarios for urban WM interesting technological and energy options. Unfortunately, none of these options can fully provide a safe waste disposal strategy or an appropriate resource management pattern for a full recovery of available resource potential of the unpresorted wastes. Further, concerning pollution problems, it does not appear that the scenarios can address the unsolved problems, social criticism, and point out the positive aspects. Bindel et al. (2012) represented the work system of product serves as a notorious system that emphasizes the transition of ownership from the customer to the manufacturer. The goal is not to sell the product, but instead, to increase its operation time, to manufacture more durable products, to design disassembly products, and to extend the use phase of the product as far as technically and economically possible. Pires et al. (2019) highlighted the fact that the waste prevention and reuse measures in a specific phase of a product should create more waste in subsequent life stages or the avoidance of toxic materials that are used to increase durability.

3.2 Innovative technologies

The recent technological innovations have increased pressure on manufacturing companies to think beyond the economic benefits of their processes and products and consider the environmental and social effects (Joung et al., 2013; Hussain and Mishra, 2018; Hussain and Kecili, 2019;

Hussain, 2019). Therefore, the ultimate goal for manufacturers is to promote manufacturing processes and manufactured products which can minimize the environmental impacts while maintaining economic benefits. The current situation is challenging manufacturing enterprises around the world to be competitive in the market place by developing and implementing sustainable manufacturing techniques and tools. Further, the manufacturers have engaged in the identification of the sustainability of measurement solutions, and a few effective measurement methods are already available for assessing the impacts of manufacturing on the environment and society. The various sets of indicators, evaluation criteria, and their characteristics are well described by Joung et al. (2013). The National Institute of Standards and Technology (NIST) indicator categorized the five dimensions of sustainability, such as environmental stewardship, economic growth, social well-being, technological advancement, and performance management. Further, Fig. 4 shows the top-level categorization and the first-level subcategorization (Joung et al., 2013).

Similarly, a paradigm of Industry 4.0 is a next step forward toward more sustainable industrial value creation. This step is mainly characterized as a contribution to the environmental dimension of sustainability (Stock and Seliger, 2016). This paradigm allocates the resources, including products, materials, energy, and water, which can be realized more efficiently based

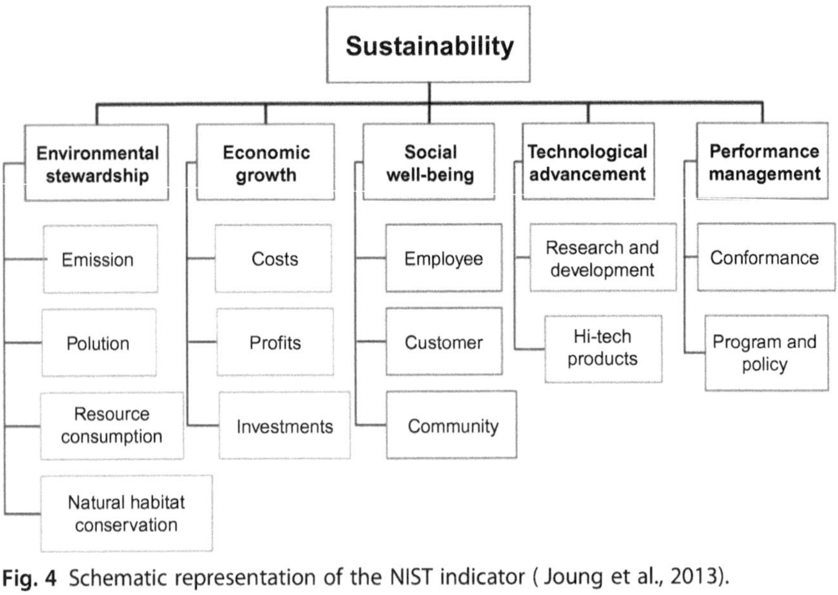

Fig. 4 Schematic representation of the NIST indicator (Joung et al., 2013).

on intelligent cross-linked value creation modules (Kagermann et al., 2015). The macroscopic perspective of Industry 4.0 is represented in Fig. 5, highlighting the end-to-end engineering cross-linking of stakeholders, products, and equipment along with the product life cycle. The life cycle consists of the raw material acquisition phase, the manufacturing phase—containing the product development, the engineering of the related manufacturing system and the manufacturing of the product, including reuse, remanufacturing, recycling, recovery, and disposal—and the transport between all phases (Stock and Seliger, 2016). Jayal et al. (2010) presented a holistic view of products and involved manufacturing processes with a particular emphasis on the supply chain, including manufacturing systems across multiple product life cycles, improved models, sustainability evaluation, and optimization techniques at the product, process, and system levels. On the same line, a critical literature review on the same has been described by Seliger et al. (2011) and Garetti and Taisch (2012). Companies are beginning to discover the implications of using additive manufacturing (AM) technologies on extending product life cycles and closing the loop.

AM, a new technology, enables the designers with nearly unlimited freedom of design, allows for mass customization of consumer goods, and offers the potential for creating such lasting objects of desire, pleasure, and attachment (Diegel et al., 2010). The AM has begun its journey for developing

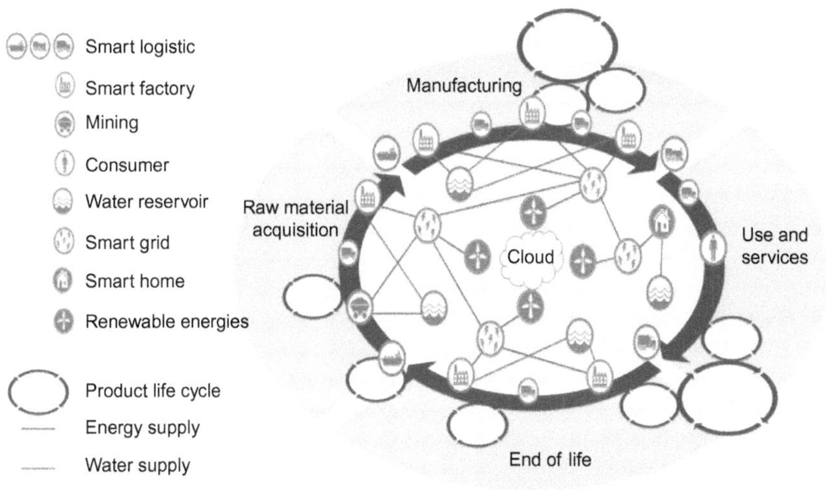

Fig. 5 Macroprospective of Industry 4.0 in the context of sustainability (Stock and Seliger, 2016).

sophisticated products for various industries (Le Bourhis et al., 2013). According to Ford and Despeisse (2016):

- The adoption of AM and other advanced manufacturing technologies appears to herald a future in which value chains are shorter, smaller, more localized, more collaborative, and offer significant sustainability benefits.
- The AM mimics biological processes by creating products layer-by-layer.
- It improves resource efficiency: improvements can be realized in both production and use phases as manufacturing processes.
- Extends product life through technical approaches through various R's.
- Reconfigured value chains into shorter and simpler.

Fig. 6 shows the AM implications in sustainability through life cycle analysis (LCA). It has been observed that the AM can provide opportunities for commercial organizations to experiment with their business models. The transition to direct digital manufacturing will lead to digital designs being kept on file remanufacturing will enable product life extension and provide incentives for product-service business models. Further, the exploitation of these avenues will lead to changes in the distribution of manufacturing and the reconfiguration of value chains. The significant changes are contingent on organizations, first for redesigning the components and the products to have fewer subcomponents subsequently leading to simplified supply chains (Ford and Despeisse, 2016).

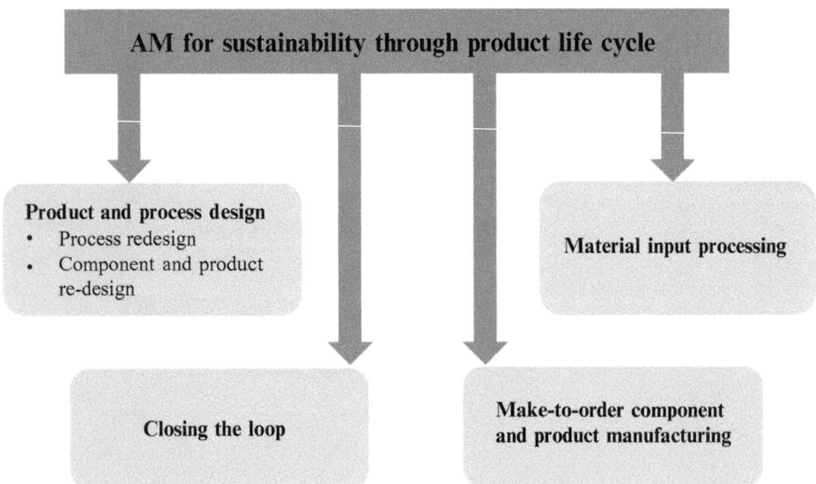

Fig. 6 Sustainability implications of AM through product life cycle.

The LCA has become one of the key techniques for studying and analyzing strategies to meet environmental challenges, for example, WM. The strengths of the LCA benefited traditional engineering and process analysis. A technological process or a change in process can produce a range of consequences whose impacts can only be perceived when this entire range is considered. The application of LCA promises to change the treatment of environmental considerations within the larger concerns of modern technological society. Recent studies are dealing with the WM through the use of life cycle assessment (LCA) modeling (Henriksen et al., 2018). The LCA of waste incineration often provides contradictory results if these local conditions are not properly accounted, and it is important to evaluate the waste incinerators processes based on different LCA models (Joyce, 2018). For more descriptive applications of LCAs in WM, the studies mentioned in Edwards et al. (2018) and San-Francisco et al. (2020) are useful.

4 Conclusions

It has been observed that the philosophical term, ZW and ZWT, of the late 20th century has now evolved as a scientific approach of controlling the waste from different sectors. Along with this, it has driven various other technological tools for the real-life estimation of the tool's effectiveness. Nowadays, Industry 4.0, AM, and LCA have been found to be key technologies that have already shown great promise to improve our understanding of the wider implications and relationships considering socioeconomic and environmental concerns while making decisions. Further, it has been found that government interferences are vital while enforcing new legislations and policies as well as subsiding sufficient funds for providing financial supports starting from the grass-root levels to the top of the administration.

References

Barnes, D., Galgani, F., Thompson, R., Barlaz, M., 2009. Accumulation and fragmentation of plastic debris in global environments. Philos. Trans. R. Soc., B 364 (1526), 1985–1998.

Bindel, A., Rosamond, E., Conway, P., West, A., 2012. Product life cycle information management in the electronics supply chain. Proc. Inst. Mech. Eng. B J. Eng. Manuf. 226 (8), 1388–1400.

Brown, L., 2008. Plan B 3.0: Mobilizing to Save Civilization. W.W. Norton & Company, New York.

Buclet, N., Godard, O. (Eds.), 2013. Municipal Waste Management in Europe: A Comparative Study in Building Regimes. In: Vol. 10. Springer Science & Business Media.

Carrico, M., Kim, V., 2014. Expanding zero-waste design practices: a discussion paper. Int. J. Fash. Des. Technol. Educ. 7 (1), 58–64.

Carroll, C., 2008, January. High tech trash. Natl. Geogr. 213 (1), 64–81.
Central Vermont Solid Waste Management District, 2009, December 18. Working Toward Zero Waste. Retrieved fromhttp://www.cvswmd.org/wp/category/working-towardzero-waste/.
Chertow, M., Ehrenfeld, J., 2012. Organizing self-organizing systems: toward a theory of industrial symbiosis. J. Ind. Ecol. 16 (1), 13–27.
Cherubini, F., Bargigli, S., Ulgiati, S., 2008. Life cycle assessment of urban waste management: energy performances and environmental impacts. The case of Rome, Italy. Waste Manag. 28 (12), 2552–2564.
Cole, C., Osmani, M., Quddus, M., Wheatley, A., Kay, K., 2014. Towards a zero waste strategy for an English local authority. Resour. Conserv. Recycl. 89, 64–75.
Connett, P., 2013. The Zero Waste Solution: Untrashing the Planet One Community at a Time. Chelsea Green Publishing.
Connett, P., Sheehan, B., 2001. A Citizen's Agenda for Zero Waste. Grass Roots and Global Video, GrassRoots Recycling Network, Canton, NY. Retrieved from http://archive.grrn.org/zerowaste/community/activist/citizens_agenda_2_read.pdf.
Couth, R., Trois, C., 2012. Sustainable waste management in Africa through CDM projects. Waste Manag. 32 (11), 2115–2125.
Danilov-Danil'yan, V., Losev, K., Reyf, I., 2009. Sustainable Development and the Limitation of Growth—Future Prospects for World Civilization. Praxis Publishing, Chichester, UK.
Diegel, O., Singamneni, S., Reay, S., Withell, A., 2010. Tools for sustainable product design: additive manufacturing. J. Sustain. Dev. 3(3). https://doi.org/10.5539/jsd.v3n3p68.
Dinshaw, A., Fortin, S., Gleason, M., Glick, D., Olivos, M., 2006. Moving Towards Zero Waste in Addison County. Middlebury College, Middlebury, VT.
Dong, H., Fujita, T., Geng, Y., Dong, L., Ohnishi, S., Sun, L., … Fujii, M., 2016. A review on eco-city evaluation methods and highlights for integration. Ecol. Indic. 60, 1184–1191.
Doppelt, B., Dowling-Wu, L., 1999. Establishing Environmentally Sustainable and Economically Efficient Economies: From Waste Management Towards Zero Waste. Center for Watershed and Community Health, Portland, OR.
Edgerly, J., Borrelli, D., 2007. Moving Towards Zero: From Waste Management to Resource Recovery. Toxics Action Center, Montpelier, VT.
Edwards, J., Othman, M., Crossin, E., Burn, S., 2018. Life cycle assessment to compare the environmental impact of seven contemporary food waste management systems. Bioresour. Technol. 248, 156–173.
Ford, S., Despeisse, M., 2016. Additive manufacturing and sustainability: an exploratory study of the advantages and challenges. J. Clean. Prod. 137, 1573–1587.
García-Ten, F.J., Quereda Vázquez, M.F., Gil Albalat, C., Chumillas Villalba, D., Zaera, V., Segura Mestre, M.C., 2016. Life ceram-zero waste in ceramic tile manufacture. In: Key Engineering Materials. Vol. 663. Trans Tech Publications Ltd, pp. 23–33.
Garetti, M., Taisch, M., 2012. Sustainable manufacturing: trends and research challenges. Prod. Plan. Control 23 (2–3), 83–104.
Gertsakis, J., Lewis, H., 2003. Sustainability and the Waste Management Hierarchy. Victoria Government, Australia Retrieved on January 30, 2008.
Girling, R., 2005. Rubbish! Dirt on Our Hands and Crisis Ahead. Eden Project Books, London.
Glavič, P., Lukman, R., 2007. Review of sustainability terms and their definitions. J. Clean. Prod. 15 (18), 1875–1885.
Grassroots Recycling Network, 2009, December 17. What Is Zero Waste? Retrieved from http://www.grrn.org/zerowaste/zerowaste_faq.html.

Hanna, M.A., 2015. From Start to Zero: The Initiation of Zero Waste Planning in Teton County, Wyoming. Doctoral Dissertation.
Havel, M., Moňok, B., Stoykova, I., Bendere, R., Tömöri, B., Popelkova, J., 2006. Zero Waste as Best Environmental Practice for Waste Management in CEE Countries. Arnika – Toxics and Waste Programme, Praha.
Henriksen, T., Astrup, T.F., Damgaard, A., 2018. Linking data choices and context specificity in life cycle assessment of waste treatment technologies: a landfill case study. J. Ind. Ecol. 22 (5), 1039–1049.
Hogland, W., Kaczala, F., Jani, Y., Hogland, M., Bhatnagar, A., 2014. Beyond the Zero Waste Concept. Linnaeus Eco-Tech.
Hussain, C.M., 2019. Handbook of Environmental Materials Management, first ed. Springer International Publishing.
Hussain, C.M., Kecili, R., 2019. Modern Environmental Analysis Techniques for Pollutants. Elsevier.
Hussain, C.M., Mishra, A.K., 2018. Nanotechnology in Environmental Science. 2 Volumes-Vol. 1. John Wiley & Sons.
Jayal, A.D., Badurdeen, F., Dillon Jr., O.W., Jawahir, I.S., 2010. Sustainable manufacturing: modeling and optimization challenges at the product, process and system levels. CIRP J. Manuf. Sci. Technol. 2 (3), 144–152.
Jiao, W., Boons, F., 2017. Policy durability of circular economy in China: a process analysis of policy translation. Resour. Conserv. Recycl. 117, 12–24.
Joung, C.B., Carrell, J., Sarkar, P., Feng, S.C., 2013. Categorization of indicators for sustainable manufacturing. Ecol. Indic. 24, 148–157.
Joyce, P.J., 2018. Environmental Considerations in the Zero-Waste Valorisation of Bauxite Residue: A Life Cycle Perspective. Doctoral dissertationKTH Royal Institute of Technology.
Kagermann, H., Lukas, W., Wahlster, W., 2015. Abschotten ist keine Alternative. VDI nachrichten. 16.
Karani, P., Jewasikiewitz, S., 2007. Waste management and sustainable development in South Africa. Environ. Dev. Sustain. 9 (2), 163–185.
Kerdlap, P., Low, J.S.C., Ramakrishna, S., 2019. Zero waste manufacturing: a framework and review of technology, research, and implementation barriers for enabling a circular economy transition in Singapore. Resour. Conserv. Recycl. 151, 104438.
Koryakov, A.G., Kulikov, M.V., 2018. Features of the development of Russian garbage-processing clusters through the implementation of project management tools. Econ. Innov. Manag. (4), 16–27. https://doi.org/10.26730/2587-5574-2018-4-16-26.
Kozlowski Russell, J., 2009. From Solid Waste Management to Solid Waste Avoidance: A Critical Evaluation of Zero Waste Strategic Plans. Master's thesisUniversity of New York, Buffalo.
Lazarevic, D., Buclet, N., Brandt, N., 2010. The influence of the waste hierarchy in shaping European waste management: the case of plastic waste. Regional Develop. Dialog. 31 (2), 124–148.
Le Bourhis, F., Kerbrat, O., Hascoet, J.Y., Mognol, P., 2013. Sustainable manufacturing: evaluation and modeling of environmental impacts in additive manufacturing. Int. J. Adv. Manuf. Technol. 69 (9–12), 1927–1939.
Lehmann, S., 2010. Resource recovery and materials flow in the city: zero waste and sustainable consumption as paradigms in urban development. Sustain. Develop. Law Pol. 11, 28.
Lehmann, S., 2011a. Resource recovery and materials flow in the city: zero waste and sustainable consumption as paradigms in urban development. Sustain. Develop. Law Pol. 11 (1), 13 28–38 and 66–68.

Lehmann, S., 2011b. Optimizing urban material flows and waste streams in urban development through principles of zero waste and sustainable consumption. Sustainability 3 (1), 155–183.

Levin, K., Cashore, B., Bernstein, S., Auld, G., 2012. Overcoming the tragedy of super wicked problems: constraining our future selves to ameliorate global climate change. Policy. Sci. 45 (2), 123–152.

Lottermoser, B.G., 2011. Recycling, reuse and rehabilitation of mine wastes. Elements 7 (6), 405–410.

Matete, N., Trois, C., 2008. Towards zero waste in emerging countries—a South African experience. Waste Manag. 28 (8), 1480–1492.

Meadows, D., Randers, J., Meadows, D., 2005. Limits to Growth: The 30-Year Update. Earthscan Publications, London.

Murray, R., 2002. Zero Waste. Greenpeace Environmental Trust, London.

Murray, A., Skene, K., Haynes, K., 2017. The circular economy: an interdisciplinary exploration of the concept and application in a global context. J. Bus. Ethics 140 (3), 369–380.

Navia, R., Ross, D., 2009. Sanitary landfills, foundation of the waste hierarchy inverted pyramid. Waste Manag. Res. 27, 407–408. https://doi.org/10.1177/0734242X09342122.

Ohrn-Mcdaniel, L., 2014. Zero-Waste Pattern Meets Technology for Marketable & Sustainable Design. Iowa State University.

Palmer, P., 2007, March 29. The death of recycling. In: Rachel's Democracy and Health News.900, Retrieved from http://www.precaution.org/lib/07/ht070329.htm#The_Death_of_Recycling.

Pariatamby, A., Tanaka, M., 2014. Municipal Solid Waste Management in Asia and the Pacific Islands. Environmental Science, Springer, Singapore.

Pires, A., Martinho, G., 2019. Waste hierarchy index for circular economy in waste management. Waste Manag. 95, 298–305.

Pires, A., Martinho, G., Rodrigues, S., Gomes, M.I., 2019. Prevention and reuse: waste hierarchy steps before waste collection. In: Sustainable Solid Waste Collection and Management. Springer, Cham, pp. 13–23.

Puckett, J., Byster, L., Westervelt, S., Gutierrez, R., Davis, S., Hussain, A., Dutta, M., 2002. Exporting Harm: The High-Tech Trashing of Asia. The Basel Action Network, Seattle.

Rios, L., Moore, C., Jones, P., 2007. Persistent organic pollutants carried by synthetic polymers in the ocean environment. Mar. Pollut. Bull. 54 (8), 1230–1237.

Royte, E., 2005. Garbage Land: On the Secret Trail of Trash. Little, Brown and Company, New York.

San-Francisco, A., Sopelana, N., Fernandez, J.M., Otegi, J.R., Minguez, R., 2020. Analysis about the incorporation of remanufacturing concept into life cycle assessment theories. J. Remanuf. 10, 1–13.

Schneider, A., 2015. Reflexivity in sustainability accounting and management: transcending the economic focus of corporate sustainability. J. Bus. Ethics 127 (3), 525–536.

Seliger, G., Khraisheh, M.M., Jawahir, I.S. (Eds.), 2011. Advances in Sustainable Manufacturing. Springer Science & Business Media.

Snow, W., 21 April 2011. Former Director, Zero Waste New Zealand Trust. Personal communication.

Stock, T., Seliger, G., 2016. Opportunities of sustainable manufacturing in industry 4.0. Procedia Cirp 40, 536–541.

Tyulenev, M., Garina, E., Khoreshok, A., Litvin, O., Litvin, Y., Maliukhina, E., 2017. A method of effective quarry water purifying using artificial filtering arrays. IOP Conf. Ser.: Earth Environ. Sci. 50 (1), 012035.

Tyulenev, M., Markov, S., Cehlar, M., Zhironkin, S., Gasanov, M., 2018. The model of direct dumping technology implementation for open pit coal mining by high benches. Acta Montan. Slovaca. 23(4).

United Nations Environment Programme, 2011. What Is Waste—A Multitude of Approaches and Definitions? Retrieved from http://www.grida.no/publications/vg/waste/page/2853.aspx.
United States Environmental Protection Agency, 2011. Municipal Solid Waste Generation, Recycling, and Disposal in the United States: Facts and Figures for 2010. Retrieved from http://www.epa.gov/osw/nonhaz/municipal/pubs/msw_2010_rev_factsheet.pdf.
Upadhyaya, L., 2013. Zero Waste. Centria University of Applied Sciences.
Van Ewijk, S., Stegemann, J.A., 2016. Limitations of the waste hierarchy for achieving absolute reductions in material throughput. J. Clean. Prod. 132, 122–128.
Veleva, V., Bodkin, G., Todorova, S., 2017. The need for better measurement and employee engagement to advance a circular economy: Lessons from Biogen's "zero waste" journey. J. Clean. Prod. 154, 517–529.
Watson, M., 2009. Waste management. In: Kitchen, R., Thrift, N. (Eds.), International Encyclopedia of Human Geography. Elsevier, London, pp. 195–200.
WCED (World Commission on Environment and Development), 1987. Our Common Future. Oxford University Press, Oxford.
Weisman, A., 2007. The World Without Us. Thomas Dunne Books/St. Martin's Press, New York.
Williams, I., 2015. Hierarchy history. CIWM J. 2015 (February), 20–22.
Wilson, D.C., Rodic, L., Modak, P., Soos, R., Carpintero, A., Velis, K., Simonett, O., 2015. Global Waste Management Outlook. UNEP.
Winans, K., Kendall, A., Deng, H., 2017. The history and current applications of the circular economy concept. Renew. Sust. Energ. Rev. 68, 825–833.
Yang, W.S., Park, J.K., Park, S.W., Seo, Y.C., 2015. Past, present and future of waste management in Korea. J. Mater. Cycles Waste Manage. 17 (2), 207–217.
Zaman, A.U., 2014. Measuring waste management performance using the 'zero waste index': the case of Adelaide, Australia. J. Clean. Prod. 66, 407–419.
Zaman, A.U., 2015. A comprehensive review of the development of zero waste management: lessons learned and guidelines. J. Clean. Prod. 91, 12–25.
Zaman, A.U., 2017. A strategic framework for working toward zero waste societies based on perceptions surveys. Theatr. Rec. 2 (1), 1.
Zero Waste International Alliance, 2009, December 18. Global Principles for Zero Waste Communities. Retrieved from http://zwia.org/joomla/index.php?option=com_content&view=article&id=10&Itemid=8#4.

CHAPTER TWO

Zero waste certification

Kailash Shweta Pal, S. Subhashini, and Kantha Deivi Arunachalam
Department of Biotechnology, School of Bioengineering, SRM Institute of Science and Technology, Kattankulathur, Tamil Nadu, India

Contents

1 Introduction 24
2 What is zero waste? 26
3 What is the zero waste strategy? 27
4 What is zero waste certification? 27
5 Why ZWC is to be done? 28
6 What are the benefits of ZWC? 30
7 What are the standard for certification? 31
  7.1 How do I choose a certification company? 31
  7.2 What items are in your waste stream, and would they be easy or difficult to recycle or reuse? 31
  7.3 Who are the companies which are certified? 32
  7.4 What is the process for certification once I have chosen a verifier? 32
8 Who certifies zero waste? 33
  8.1 Zero waste international allowance 33
  8.2 Gbci's true zero waste 35
  8.3 Ul environment certification 36
  8.4 Green circle 37
9 Zero waste certificate initiatives around the world 37
  9.1 Canberra, Australia 37
  9.2 New Zealand 38
  9.3 Nova Scotia, Canada 38
  9.4 Prince Edward Island, Canada 39
  9.5 Communities near Toronto, Canada 39
  9.6 Boulder, CO 39
  9.7 Communities in California 39
  9.8 Italy 40
10 Conclusion 41
References 41

Concepts of Advanced Zero Waste Tools
https://doi.org/10.1016/B978-0-12-822183-9.00002-7

1 Introduction

Considering the growing concern with waste management, it is important to understand and maintain a zero waste (ZW) environment. To research ZW a systematic literature review was conducted, in which more than 100 published papers were reviewed with the goal to, initially, comprehend the principle of zero waste, and, then, map its advantages, obstacles, and essential success factors. The findings indicate that the scholars did not find a consensus on the ZW definition. While some studies discuss this ideology properly, other studies are based on just one or a few of its topics. The benefits have been categorized and organized into four dimensions: community benefits, financial-economic benefits, environmental benefits, and benefits to industry and stakeholders. About the obstacles, both in the macroenvironment (mainly political and cultural) and in the mesoenvironment and microenvironment (stakeholders, industries, and municipalities) were established. Articles' review allowed the listing of key success factors, accompanied by a collection of activities that needed to be conducted. Concerning future studies, it should be noted that more empirical studies are required on the implementation of ZW, in particular on educational practices designed to facilitate changes in user behavior (Pietzsch, 2016). Waste is a sign of every industrial society's inefficiency and portrayal of misallocated capital. In the urban planning cycle, waste management systems have not earned as much attention as other sectors such as water or electricity. The new strategy will also find many holes in waste management. Global climate change and its complex impacts on human life are moving today's society toward more sustainability. The depletion of scarce global resources often forces us to consider the stewardship of resources and goods (Song et al., 2015). Many industrial and metropolitan activities, including households, generate substantial amounts of solid waste all over the world every day. Even though solid waste has been recycled and used for energy use, the production of waste is still growing due to population growth. Although solid waste has been recycled and used for energy use, waste production is still on the increase as a result of population growth. Solid waste that was discarded by society daily includes many essential components such as zinc, copper, nickel, chromium, and lead. Around the same time, human is wasting large amounts of valuable metal resources, and recent scientific literature has highlighted the ongoing depletion of natural reserves (including metals), contributing to higher metal market prices (Meylan and

Spoerri, 2014). According to Boesch et al. (2014), this pattern of increase in natural resource prices is projected to be even more pronounced in the near future, not only the prices have increased on the market but also the population is rising and more than 50% of the world's population now lives in cities. People are constantly demanding material products, and no new patterns suggesting other ways of living and consuming can be found. Therefore there is a growing need to supply all the citizens with commodities in an economically and socially sustainable manner which brings with this the need to shift the traditional linear models followed by our society not only into locked-loop models but also to consider the recovery of all valuable resources that human has lost as waste products from our daily activities and through history.

There is also a critical need to aim at waste reservoirs (soils, sludge, slag and incineration ash, bottom sediments, landfill soils, mining waste, etc.) that are completely polluted with metals and nutrients as potential secondary stocks of these important components. This leads us to the topic of the paper and the implementation of the "Beyond the zero waste principle" which encourages the recovery of all materials lost throughout the entire life cycle of various manufactured products that are still available in various sinks (landfills, river sediments, oceans, etc.). By theory, all waste, materials and chemical substances lost as sludge, slag, harbor sediments and others should return to the anthropogenic loops and remove harmful substances from the systems and treat them in an environmental- friendly manner. The long-term goal is to incorporate such a revolutionary approach in an environmentally and economically productive manner, using the accumulated expertise, including the reuse/recycling of materials in urban and rural structures. Including landfill mining, glass mining, harbor and bay mining, and seafloor mining. The zero waste principle was initially implemented as a way of minimizing the amount of solid waste that ends up in landfills in favor of integrated solid waste management systems that result in incineration. Measures were introduced to eliminate toxic substances in solid waste and ban imports; mercury thermometers were manufactured and sold in 1993–94, a leading source of mercury in Sweden's waste in the late 1980s. To improve recycling and reuse, since 1994 all municipal solid waste has to be sorted before final treatment and a higher number of different sorting schemes have been implemented. The definition of eco-efficiency has also been implemented, referring to an environmentally oriented strategy aimed at preventing and minimizing the adverse effect of goods on the environment and human health over their life cycle. The idea was also to improve the quality of waste fuel resulting in incineration while improving the incineration technique and the treatment of flue gas when the EU Council Directive 99/31/EC of 1999, which was subsequently followed late in 2008/98/EC of 2008 on EU landfilling, countries were obliged to reduce biologically degradable waste resulting in landfills. This encouraged incineration and the building of large numbers of incinerators, and the number rose over 15 years by more than 60%. During this time the need for the temporary storage of waste fuels increased and the risk of self-ignition fires also increased in the time 2000–2008, as there was not a sufficiently large incineration capacity among the incinerators that time (Hogland et al., 2019).

2 What is zero waste?

Zero waste is a set of standards based on waste prevention, which promotes the resource life cycles to be revamped to reuse all items. The goal is not to send any garbage to landfills, incinerators, or the ocean. Just 9% of plastic is recycled. The material will be reused in a zero waste program before the optimum usage level is reached. Zero waste contains and uses products that would be dumped into nature and the oceans without producing them (Song et al., 2015). Zero waste is a lifestyle, a way of thinking, and a philosophy for the company. For example, never purchase, use, or supply plastic straws, plastic bottles, or packaging! You also help to clean up nature and the oceans by saying no to plastics (Zero Waste Manifesto, 2018). Zero waste stops waste from being sent to landfills that pollutes nature and the environment, and it motivates businesses to produce even more responsibly. Conservation of all resources through responsible processing; use, reuse, and recycling of goods; packaging and materials without burning; and no ground, water, or air discharges that endanger the environment or human health as described by the Zero Waste International Alliance (ZWIA) (Zero Waste International Alliance, 2020).

Values to lead a zero waste life:

- Well-being of plants and animals in relation to customer needs and ease of use
- Natural resources in relation to mass production
- Recycling in relation to one-time use
- Local fresh goods in relation to global frozen items
- Use of goods until the "Best By" date in relation to smart packaging
- Cloth bags over plastic bags

It is just meant to motivate the reader to zero waste movement or inculcate zero waste idea or concept in their lives. Join any Zero Waste Project that is close to you. If your neighborhood does not have a Zero Waste Network, you can always create one, educate and motivate people, and organize about how to live a zero waste life (Hogland et al., 2019).

3 What is the zero waste strategy?

The proposal for zero waste says "no to incinerators," "no to mega-landfills," "no to the throwaway system," and "yes to a safe world." While it will sound like an idealistic goal, we can set it in a realistic time frame. Next year, we are not planning to hit zero waste, but we are aiming to close some places to zero waste by 2020. Some are going to quibble about how close we can get to zero but the idea is to make it our objective. Let me put this question another way: bearing in mind the needs of future generations, how much waste do you think is acceptable? Zero waste is more of a new approach than anything else. We have to switch to the front end of resource management and better industrial design from the rear end of waste disposal. We need to build proper method and remove waste from the dictionary. As Mary Lou Derventer, who operates one of the world's largest reuse operations along with her husband Dan Knapp, says, "Discarded materials are not waste unless it is wasted." Waste is a verb, not a noun (Hogland et al., 2019).

4 What is zero waste certification?

Zero waste means designing and maintaining goods and procedures to actively prevent and remove waste and material quantity and contamination, conserving, and preserving all resources, and not burning or burying them. The implementation of zero waste would eliminate all land, water, or air discharges that pose a threat to planetary, human, animal, or plant health (Hogland et al., 2019). Zero Waste Certification (ZWC) is a designation that requires the removal of at least 90% of total waste from landfills, and the introduction of other programs to minimize our environmental impact (I Heard That Your Company Is Zero Waste Certified, n.d.). The ZWC process is to check how efficiently organizations can carry out the reduction of nonhazardous waste and try to recycle. It is a long process that involves focusing on both upstream processes and policies by implementing zero waste practices in the industries. ZWC shows the skill of the waste management or recycling facility. The goal of each facility is to remove all solid, nonhazardous waste from landfill, incineration, and the atmosphere. Earning this credential results in market value, the longevity of the device, emissions control, and others. Certification is required for every physical establishment and its operations. The ZWC process assesses how effectively organizations can achieve nonhazardous solid waste reduction and use capital to maximize their production. It is a long-term process that includes concentrating on both upstream processes and policies by enforcing best practices in the industry that make zero waste possible, rather than focusing solely on production.

Most critically, facilities will have in place recycling structures that effectively incorporate zero waste products into the environmental environment. A company that pursues a ZWC must be able to meet the requirements by consistently reducing the use of raw materials and increasing production performance while minimizing environmental waste and understanding how to reduce the total waste (Home and Waste?, 2019).

5 Why ZWC is to be done?

When disposability is increasingly unfashionable, businesses are taking new steps toward zero waste activities that the ZWIA has described when activities that deliberately prevent and minimize the quantity and contamination of waste and materials, conserve and recover all resources and do not destroy or bury them. Though zero waste is becoming a prominent lifestyle choice for households, its utility is only starting to enter the psyche of the corporate world (US EPA, 2020).

In September 2017, the Green Business Certification Incorporation, most popularly known for its "LEED" certification scheme, launched the "TRUE" ZWC program. This program works similarly to LEED's point system for certification, which measures buildings' environmental performance. Some requirements for certification with TRUE include achieving at least 90% diversion from landfill, incineration, and the environment for solid, nonhazardous wastes for the most recent year as well as ensuring that materials do not exceed a 10% contamination level at the point of disposal. TRUE provides workshops and trainings for business professionals to make the certification process easier and more accessible. TRUE has already certified several businesses across the United States and Canada, including the Sierra Nevada Brewing Company in North Carolina, the Yellowstone General Stores Warehouse in Montana, and, unsurprisingly, the Tesla Factory in California (Zerowaste Design, 2020).

According to the Environmental Protection Agency's 2016 Recycling Economic Information Report, zero waste practices such as recycling and composting are economically beneficial. Recycling, on average, creates 10 times as many jobs as trash, and composting creates twice as many jobs as landfills. You Sow, an advocacy group, determined that 11 billion dollars is wasted every year by sending recyclable materials to landfills. One success story involves the automotive company Subaru. Since eliminating all of their landfill waste, both its American (in Indiana) and Japanese plants have saved between 1 and 2 million dollars per year by reusing materials. These statistics reveal that zero waste is an asset, not a far-off environmentalist dream (Northeast Recycling Council, 2020b).

However, zero waste consulting firms are quick to point out the distinction between "zero waste" and "zero waste to landfill." The latter term accepts incineration as a way of halting inputs to landfills. Designating businesses as "zero waste to landfill" is often a greenwashing tactic that hides the negative effects of waste-to-energy technology. The only marketable waste-to-energy technology right now is mass burn, which emits excessive amounts of carbon dioxide and harmful air pollutants. Additionally, waste-to-energy technology does not provide as many jobs and cost-saving benefits as recycling and composting. While the Subaru Indiana plant has not sent any materials to landfills since 2004, less than 5% of its nonhazardous waste is incinerated to generate energy. The ashes are then used for road-resurfacing materials. On the other hand, some companies practice waste-to-energy as their primary mode of "preventing" trash (Northeast Recycling Council, 2020c).

While some argue that the environmental impact of the industry must be addressed before the environmental impact of households, the two are heavily intertwined. Certification programs for Fair Trade and Rainforest Alliance products work because consumers have demonstrated a willingness to pay for these products. Consumers in developed countries, however, are often connected to develop markets and the rainforest only by purchasing certified products. Zero waste requires the action of both the producer and the consumer for it to be successful, so consumers must express demand for zero waste products for certification to have any genuine significance. Our landfill issue will not be solved if consumers support businesses that produce significant amounts of waste (Google.com, 2020).

The only way everyday consumers can help to limit the amount of waste sent to incineration sites or landfills is by actively reducing the amount of waste in all spheres of their lives. Subaru, for example, made sure that

96% of the components in their vehicles can be recycled or reused by the consumer. But this does not incentivize zero waste practices for their customers. A desire for consumers to reduce waste can only be instilled through education and awareness of waste streams and their disastrous effects on ecosystems and public health. Nevertheless, the infiltration of the zero waste ethos to the business world represents a positive shift not only for the environment but also for the economy. Investing in ZWC programs means investing in new jobs, local recycling and composting facilities, and novel ways of repurposing what is now known as trash (Gbci.org, 2020a).

6 What are the benefits of ZWC?

Establishing zero waste policies also helps companies to turn waste into income, as rising waste contributes to cost savings in the long term. Such organizations will minimize carbon emissions, minimize trivial waste and pollution and generate sustainable value for the world around them. Organizations that pursue ZWC are required to divert a certain percentage (sometimes up to 90%) of their waste away from the surrounding environment, landfills, or incinerators (Home and Waste?, 2019).

In crux, the following advantages come with receiving a ZWC:

1. Facilitates in mitigating air, water, and land pollution that are detrimental to both health and the surrounding atmosphere.
2. It helps to reduce overall costs and thus, improves net profitability.
3. Increases the use of recycled materials which lessens an organization's reliance on polluting supplies and reduces its overall carbon footprint.
4. Enhances an organization's brand value by showcasing its responsibility and commitment to the local society and environment (Home and Waste?, 2019).

Governments frequently set industry waste management goals and occasionally even provide incentives. Most governments even implement waste management projects at a local level, utilizing shared regional-level facilities.

To achieve the zero waste target, manufacturing companies and product designers will concentrate on creating goods that can be easily broken down for recycling, can be reused industrially in the production of any other product, or can be decomposed naturally by the external environment (biodegradable).

Additionally, the products should be durable and repairable, which increases their usable life cycle and lessens the unnecessary waste of resources and raw materials. Manufacturers and designers should also focus on minimal or environmental friendly packaging which does not add waste after serving its purpose.

The definition of "zero waste hierarchy" describes a hierarchy of strategies and initiatives to maintain the zero waste program, ranging from the most desirable uses of resources to the least preferred. Structured in a manner that can be applied to anyone from law-makers to businesses and everyday folk, the aim of the concept is to provide a deeper insight into the internationally popular 3R's (Reduce, Reuse, and Recycle) and has, to a certain extent, been incorporated into recycling regulations and resource conservation policies across the globe.

The zero waste hierarchy also promotes policymaking, enforcement, and fundraising or expenditure at the height of the hierarchy, and offers advice to those who wish to adopt waste disposal processes or manufacture products that get them as closely as possible to zero waste. It also offers ways and methods to strategize and evaluate resources (Home and Waste?, 2019).

7 What are the standard for certification?

There are no universal standards for becoming zero landfill certified. The standards that you will have to meet are determined by the third-party verifier you choose. Even though a verifier's certification criteria is not universally recognized, meeting their standards adds credibility to your claims and proves to your stakeholders that you are in fact doing what you are claiming.

7.1 How do I choose a certification company?

Selecting the third-party verifier is probably the most important step in the process because you are choosing both a verifier and a standard in the same selection. Clearly defining your goals for waste solutions ensures that you are choosing a verifier that validates those goals and helps you to achieve them. Choosing the wrong company will hinder your organization's ability to reach goals and become certified. North Star Recycling offers some questions companies need to answer for themselves before reaching out to a zero landfill verification company.

7.2 What items are in your waste stream, and would they be easy or difficult to recycle or reuse?

Is diversion of waste a key priority for your company, and have you set hard goals?

- Is your company targeting the entire supply chain or only disposal practices?
- Are there any restrictions your company wants to be enforced for disposal practices?

Once you have the answers, you can begin to consider certification companies and their individual standards. Covanta strongly suggests doing your due diligence on the front end of the search for a verifier to avoid any surprises. The most important task is to be sure that their standards align with your organization's goals. Also, consider the reputation of the verifier and whether they have ever worked in your industry. Those who have experience in the same industry may be able to offer insights for improvement and identify any risks the company may be exposed to. Finally, it helps to get to know the team from the verifier with whom you will be working. Establishing a strong foundation will help the process to go smoothly.

7.3 Who are the companies which are certified?

There are a variety of organizations to choose for the certification process. Some of the businesses with the most exposure are as follows:

1. Green Business Certification, Inc. (GBCI)
2. Intertek
3. Underwriter's Laboratory (UL)
4. Green Circle Certified
5. The Carbon Trust
6. NSF International

7.4 What is the process for certification once I have chosen a verifier?

Covanta goes on to explain that no matter which third-party certification company you choose, the process for certification will be nearly the same. It usually begins with a site visit in which auditors look at what waste streams are being generated and how they are being managed. They will audit the procedures for waste management and training programs that are in place for employees. Auditors will also do an intensive review of all records and documentation that demonstrate where waste has gone and how it is managed at its final destination. They may also visit downstream waste outlets to verify your claims. Once they have completed their audits, they will either grant you certification or leave you with recommendations for improving your processes to meet their certification standards (Gbci.org, 2020b).

8 Who certifies zero waste?

8.1 Zero waste international allowance

8.1.1 Benefits of ZWIA zero waste community recognition

National Affiliate is authorized to publicly say "The Zero Waste International Alliance has reviewed data provided by the community and determined that they have adopted a zero waste goal and are either working towards or diverting over 90% of their discards from landfilling or incineration according to ZWIA Global zero waste community principles." May be listed as 1 of 2 categories: "Communities Working toward Zero Waste" or a "Zero Waste Best Practice Community" on ZWIA website and website of ZWIA National Affiliate (Gbci.org, 2020a).

8.1.2 Minimum criteria that ZWIA recognizes as a zero waste group

i. Adopted goal of zero waste that uses ZWIA definition of zero waste as summarized here:
   - **(a)** All discarded materials are resources.
   - **(b)** Resources should not be burned or buried.
   - **(c)** Goal is zero air, water, and land emissions.

ii. Meet all national, state/provincial, and local solid waste and recycling laws and regulations.
   - **(a)** Submit summary of their zero waste initiatives that can be published on ZWIA and National Affiliate websites and indicate the official title of their agency.
   - **(b)** Submit data annually to National Affiliate and demonstrate progress in implementing zero waste plan or strategy. With these annual renewals a whole year of data will be given. The data submitted will be made public and will be reported on the website of the National Affiliate (Gbci.org, 2020a).

8.1.3 Categories of recognition

Communities working toward zero waste the group must be remembered as a society working toward zero waste:

i. Adopt a commitment to implement residential collection programs for recyclables and organics (including food scraps) by a given date.

ii. Consider all discards generated in the community whether they are directly controlled by the community (such as discards generated in the institutional, commercial, and industrial sectors).

iii. Communities should exercise control over those sectors they are directly responsible for and influence those sectors that they are not directly responsible for.

iv. Advocate for redesign of problem materials that are not recyclable or compostable. Consider local actions/campaigns to encourage redesigns.

v. Report progress annually toward zero waste plan milestones.

(a) Implement a pay-as-you-throw rate structure or other financial incentives for generators (if allowed by state/provincial or national regulations) to encourage them to waste less and recycle more.

(b) Establish a zero waste advisory board or multi-stakeholder mechanism to engage in the invention and execution of a zero waste plan or strategy, identify critical measures, identify workarounds or reset deadlines, and develop a specific policy, system, and encourage the implementation of a zero waste plan or decision-making strategy.

(c) Perform detailed compositional analyses of discarded materials at least every 10 years to evaluate the progress of the zero waste program, determine what remains of discarded materials, identify strategies and programs for further development, provide input to manufacturers, and work with them on the redesign of hardly reusable, recyclable materials, goods, and packaging. To satisfy this criterion, equivalent data or analysis of product and service opportunities can be used. Consider more comprehensive annual evaluations of residual resources to track development.

(d) Oppose any form of incineration (technologies above 212°F or 100°C), both those already operating (legacy incinerators) and those planning or developing within their jurisdiction or region. Communities with current incinerators must agree in writing to phase out all incineration in the next contract with service providers or when there are alternatives accessible.

(e) Defining mid-term and long-term quantitative targets; this may include a residual waste reduction goal or a further reduction within 10 years, or the introduction of "darn near Zero."

Such activities should be included either in a formal zero waste resolution and/or a zero waste plan or strategy signed by the individual with jurisdictional authority (major, manager, council, district, or otherwise, depending on the local regulatory system and the local parties' established responsibilities) (Gbci.org, 2020a).

Zero waste best practice communities: These are communities that follow the guidelines for communities working toward zero waste and that demonstrate best practices and actual achievements on the road to zero waste.

For those sectors under their direct jurisdiction, there are four levels of recognition for communities that reach various rates of diversion of all discarded material:

i. Achieved 50% diversion from landfills, incinerators, and the environment.
ii. Achieved 70% diversion from landfills, incinerators, and the environment.
iii. Achieved 90% diversion from landfills, incinerators, and the environment.
iv. No burn and diverted 90% from landfills and the environment (Gbci.org, 2020a).

8.2 GBCI's true zero waste

The TRUE ZWC program of GBCI is used by facilities to identify, meet, and attain their zero waste goals, to reduce their carbon footprint, and to support public health. The certification goes beyond the number of diversions and focuses on the upstream policies and practices that make zero waste in any organization and beyond successful. TRUE-certified spaces are environmental friendly, more resource effective and help to turn waste into savings and extra revenue streams. They cut greenhouse gases by closing the loop, manage risk, minimize litter and waste, reinvest capital locally, build jobs, and add value to their organization and community.

TRUE-certified institutions.

- Save money: Waste is a symbol of inefficiency and waste management is cost savings.
  - Greater progress: a zero waste policy strengthens manufacturing practices and environmental management measures, leading to broader, more ambitious actions.
  - Sustainability support: The zero waste approach supports the three P's—people, planet, and profit.
  - Boost resource flows: A zero waste policy uses fewer fresh raw materials and sends no waste products to landfills, waste-to-energy incineration, and the atmosphere.
- Implementing TRUE ZWC criteria and credits has the following benefits for any facility pursuing a sustainable approach to resource utilization and facilities operations:
  - Helps to eliminate pollution—in our air, water, and land—that threatens public health and ecosystems.
  - Improves their bottom line by cutting costs.

- ° Reduces the ecological footprint by reducing materials, using recycled and more benign materials, and providing longer lives for products by increasing reparability and end-of-life disassembly.
- ° Promotes positive environmental and economic growth forces in the built environment by protecting the environment, raising prices, promoting the creation of new markets, and creating employment throughout our economy.
- ° Promotes active overall engagement through training of all staff and zero waste partnerships with suppliers and customers.
- ° Allows the facility to display its residues (Home and Waste?, 2020).

8.3 UL environment certification

After 1894 UL have made the planet a healthier, more prosperous environment. Buyers, building specifications, and buying companies around the world appreciate their Environmental and Sustainability Certifications. On acquiring one of our Labels, customers can be confident that the statements made regarding these goods have been checked and confirmed to the highest degree. That is how they help you to develop trust and success in the marketplace and that is what distinguishes you from the pack (Control Solutions Inc, 2020).

Environmental Product Certifications tell the story of the environmental performance of a commodity throughout its lifecycle, helping consumers to recognize holistically greener goods and achieve their sustainability objectives. Such voluntary, multiattribute, life cycle-based environmental certifications suggest that a product has undergone extensive scientific testing, comprehensive auditing, or both, to confirm its compliance with strict environmental performance requirements by third parties. Such standards set benchmarks for a wide range of parameters including: reduction of energy, recycling of waste, recyclability, and use of salvaged materials, protection of the environment, reduction of transportation, and conservation of natural resources. UL has developed multiattribute sustainability norms for the following industries: construction materials/finishing, household cleaning, toys, mobile phones, and office supplies (Intertek.com, 2020).

UL Environment sets guidelines for the quality and sustainability of a wide range of building products, consumer products, and organizations. Sustainability standards are developed through a collaborative process, which adopts a holistic approach to the creation of standards (The Emerald Review, 2020).

8.4 Green circle

Green Circle will validate and certify zero waste or a percentage of waste a company or manufacturing facility diverts from landfill. After performing a material flow and mass flow analysis, Green Circle will be able to quantify and certify a percentage of waste that has been diverted from landfills.

Objective: Demonstrate an organization's commitment to responsible management of end-of-life materials by quantifying and certifying total diverted waste.

Definitions: Waste Diversion: The prevention and reduction of generated waste through source reduction, recycling, reuse, or composting.

Source reduction: A method that removes waste at source in the first place by not producing it. This may include product and packaging redesign, process capacity changes, stock substitution, inventory management, improved housekeeping, and/or preventive maintenance.

Reusable: Products that are designed to be and can be used more than once.

Recycling: A process which transforms materials, which otherwise would be waste, into valuable resources.

Alternative daily cover: Material other than earthen material placed at the end of each operating day on the surface of the active face of a municipal solid waste landfill to control vectors, fires, odors, blowing litter, and scavenging.

Energy recovery: The conversion of nonrecoverable materials into heat, energy, or fuel through a variety of methods, including combustion, gasification, pyrolysis, anaerobic digestion, and recovery of landfill gas, is made from waste. This is often referred to as waste-to-energy (ACT, n.d.-a).

9 Zero waste certificate initiatives around the world

9.1 Canberra, Australia

In 1996, Canberra became the world's first city to implement a zero waste law (ACT, n.d.-a). The legislation required the government to generate "No Waste by 2010." In 2004, when I last visited the city, there was more than 70% diversion but to be fair this figure was primarily affected by the massive diversion of both yard waste and (very heavy) construction and building debris (ACT, n.d.-b). The most promising aspect of their plan was the creation of a "Resource Recovery Park" in an attempt to co-locate all the businesses that can produce goods from different resources and certain recycled marketing items such as "Aussie Junk" and "Revolve." Unfortunately, several delays have arisen late in the program (ACT, n.d.-c).

9.2 New Zealand

About 70% of societies have adopted a zero waste plan, a policy that the national government has now endorsed (Sound Resources Management Group, Inc, 1992).

9.3 Nova Scotia, Canada

Halifax tried to expand its groundfill in the late mid-1990s. It created a major outcry from local residents who angrily protested about the awful odors coming from the site. The municipality instead introduced a massive garbage incinerator (750 t per day). Again an outcry occurred, and the proposal was rejected. The government handed over the issue to the public at this point and said, "You don't want landfills and you don't want incinerators, tell us what you want. You design the program" (Videotape, n.d.; Nova Scotia program, n.d.-a, Nova Scotia program, n.d.-b).

The people welcomed the challenge and all the consultants' reports were issued by the Government. The people choose one of the programs listed in the Seattle-based report prepared by Sound Resources (New Zealand, n.d.). This strategy involved separating the sources and gathering recyclables, organics, and residuals from door-to-door. The people made two modifications to the text. (1) Whenever the study used the word "waste," it was changed to "capital." (2) Because of the bitter experience of the old landfill, the people requested that no organic waste would enter the landfill without processing. This has led to the development of the earlier-listed residual screening facilities (Videotape, n.d.; Nova Scotia program, n.d.-a, Nova Scotia program, n.d.-b).

This program proved to be extremely popular and successful. The entire province had reached a 50% diversion rate after 5 years—and became the first province to do so in Canada. Halifax reached a diversion rate of 60% (Videotape, n.d.). Collection and disposal of the discarded materials were produced during the course of doing these 1000 jobs. These jobs were created in materials recovery facilities, a tyre recycling plant, a waste paint recovery plant, deconstruction and reuse operations, and drop-off facilities for hazardous products such as batteries and solvents, composting facilities, the approximately 200 "eco depots" that recycle the containers, with a deposit on them, administration and study (Videotape, n.d.). In addition, a further 2000 jobs were produced in the industries that reused the collected content. In Nova Scotia's own industries, nearly all the separate products are reused. The program was the subject of a genuine progress index analysis which includes estimated social benefits unlike other indices such as gross national product. The conclusion of this study is largely optimistic due to the social implications of so many new jobs being generated (GPIAtlantic, n.d.). It is worth remembering at this point that only 80 full-time jobs were created by the much-vaunted trash incinerator in Brescia, Italy, which cost 300 million Euros to construct!

9.4 Prince Edward Island, Canada

This Canadian community has door-to-door recyclables and compostable obtained from every household on the island ((Prince Edward) Island Waste Management Corporation, n.d.).

9.5 Communities near Toronto, Canada

Markham City (north of Toronto) has in 2 years diverted 70% from landfill (Markham, Ontario, Canada, n.d.). In a pilot project, Ward 3 in Pickering, Ontario, achieved a 73% diversion rate with recyclable and compostable door collection.

9.6 Boulder, CO

Eco-cycle has an incredibly long history of recycling in Boulder, CO. One of the gurus of the zero waste campaign is their new CEO, Eric Lombardi. Its website is a mine of theoretical to practical knowledge (Ecocycle, Boulder, Colorado, n.d.).

9.7 Communities in California

Thousands of towns and cities have accomplished this goal since the enactment of a state law requiring municipalities to divert 50% of their waste from landfills by the year 2000 (Waste diversion in California, n.d.). Many of the more informed civic bodies asked after doing so, "Why stop at 50%, why not 60% or 70% or higher? Should not take a 100% shoot?" A zero waste plan has been announced by both Del Norte, Almeida, and many other counties (Zero Waste International Alliance, n.d.). Most impressively, the city of San Francisco, with 850,000 inhabitants and very little land, is among the cities that have embraced a zero waste goal by 2020. They currently divert over 60%, and their 2010 goal is 75% (Gary Liss, n.d.; Goodyear, 2007).

9.8 Italy

Italy has developed some of the world's most cost-effective and fast-tracked "door-to-door" collection systems. The program began when farmers approached Enzo Favorino and his colleagues at the Parco Monza Agricultural School near Milan and asked where they could acquire more organic material for their land. Enzo responded that in the domestic waste stream there was plenty of organic material but that it could only yield compost that was strong enough to be used in agriculture if it was collected at source separately. It will involve collecting door-to-door. Thus the very famous selection scheme "porta a porta" in Italy began (Favoino, n.d.). More than 1000 communities in Italy are now achieving more than 50% diversion using porta collection systems as of the beginning of 2008 (Favoino, n.d.). Several small towns (e.g., Sermonetta) in the province of Lazio produced more than 60% diversion in 1 year (Favoino, n.d.) In Novara, a town close to Turin with a population of 100,000, 70% diversion was accomplished in just 18 months (Novara, Italy, n.d.). To the best of our knowledge, it is a world record for a city of this size to reach such a diversion rate in such a short time. Small towns around Salerno have reached a diversion of more than 70% (Favoino, n.d.). Twenty-two communities in the Priula Consortium have reached an average diversion of 76% in the Treviso region in 5 years, with four towns over 80%. This program's economics has been closely tracked and separation programs are currently cheaper (Euros 74 per tonne) relative to nonseparation programs (Euros 93 per tonne) (Treviso region, Italy, n.d.). An 83% diversion has been reached in Villafranca d'Asti (population 30,000) in the province of Piedmont (Villa d'Asti, Piedmont, Italy, n.d.). Capannori near Lucca on February 24, 2007, became the first city in Italy to announce officially a zero waste plan. A huge 83% diversion is accomplished in a pilot port, a prota collection network, for one quarter of the city. In addition, the remaining 17%—the residual fraction—is analyzed. In this proportion, the top three products are as follows: (1) clothing and textiles; (2) disposable diapers, and (3) kitchen waste. They have already introduced reusable diapers in their supermarkets, raising their grab rate for kitchen waste and finding local leather and textile uses. The top three items in this proportion are as follows: (1) clothes and textiles; (2) disposable diapers; and (3) kitchen scrap. Through their supermarkets they have already implemented reusable diapers, increasing their grab rate for kitchen waste and seeking local uses of leather and textiles (San Francisco Zero waste program see videotape, n.d.).

10 Conclusion

Zero waste program, a total waste stream of 566 t/year, and a 25% diversion rate are main indicators for enhancing the future. Specific goals for waste reduction should be set, and progress should be assessed annually (or more often). Given a two-prong approach to achieve the target—reduction and diversion—it will be necessary to continue to calculate and evaluate absolute quantities. It is possible that the percentage of waste diverted to landfill will fluctuate up or down, thus reducing the total amount of waste produced at the facility (Gbci.org, 2020a). The zero waste and the non-zero waste idea will also be created and strengthened so that more people can reach a high material quality and the lifestyle is unlikely to change dramatically in the near future (Hogland et al., 2019).

Technology has a vital role to play in transforming our community to achieve zero waste, along with its structures and the people. Technology will push us to adjust our actions and consumption habits so that we can may, distribute, reuse, and allocate space for discards' management so that they can also be reused or recycled (Home and Waste?, 2020).

1. Zero waste may be an alternative idea in Indonesia's waste management, since zero waste is a philosophy that starts from, stops waste from "upstream" to "downstream," not just by dumping it to landfill.
2. Require the involvement of all participants, including private parties, states, and societies, in the implementation of the zero waste concept.
3. National government policy support in the form of a firmer regulation is needed if ZW is to be properly enforced (Intertek.com, 2020).

ZWC has a major role to play in regulating and handling zero waste. To improve the climate, it is therefore necessary for businesses and societies to take up the ZWC.

References

(Prince Edward) Island Waste Management Corporation n.d. http://www.gov.pe.ca/tpw/iwmc-info/index.php3.

ACT NOwaste website n.d.-a http://www.tams.act.gov.au/live/Recycling_and_Waste.

ACT NOWaste featured in the video "On the Road to Zero Waste, part 3: Canberra" produced by Paul Connett for GG video, 2003, 29 minutes, GG Video, 82 Judson Street, Canton, NY 13617, contact pconnett@gmail.com n.d.-b.

ACT n.d.-c NOwaste concerns. Contact Gerry Gillespie; gerry.gillespie@environment.nsw.gov.au.

Boesch, M., Vadenbo, C., Saner, D., Huter, C., Hellweg, S., 2014. An LCA model for waste incineration enhanced with new technologies for metal recovery and application to the case of Switzerland. Waste Manag. 34 (2), 378–389. https://doi.org/10.1016/j.wasman.2013.10.019.
Control Solutions Inc, 2020. Getting Your Facility Zero Landfill-Certified: What You Need to Know. [online]. Control Solutions Inc. Available at: https://controlyourbuilding.com/blog/entry/getting-your-facility-zero-landfill-certified-what-you-need-to-know.
Ecocycle, Boulder, Colorado, Contact Eric Lombardi http://www.ecocycle.org/Aboutus/EricLombardi.pdf n.d.
Enzo Favoino, n.d. Scuola Agraria del Parco di Monza, favoinomail@tin.it, www.monza.flora.it/compost.
Gary Liss n.d. gary@garyliss.com http://www.grrn.org/zerowaste/business/index.php and http://www.grrn.org/zerowaste/business/profiles.php and http://66.35.240.8/cgibin/article.cgi?f=/c/a/2007/07/20/BAGCSQV2661.DTL.
Gbci.org, 2020a. Are You Eligible For TRUE Zero Waste Certification? [online] Available at. GBCI. https://gbci.org/are-you-eligible-true-zero-waste-certification. *.
Gbci.org, 2020b. How to Achieve TRUE Zero Waste Certification. [online] Available at. GBCI. https://gbci.org/how-achieve-true-zero-waste-certification. *.
Goodyear, C., 2007. San Francisco first city to ban plastic shopping bags. San Francisco Chronicle. Wednesday, March 28 http://www.sfgate.com/cgibin/article.cgi?file=/c/a/2007/03/28/MNGDROT5QN1.DL.
Google.com, 2020. What Is Zero Waste Certification - Google Search [online] Available at. https://www.google.com/search?q=what+is+zero+waste+certification&oq=what+is+zero+waste+certification&aqs=chrome.69i57.10494j0j7&sourceid=chrome&ie=UTF-8. *.
GPIAtlantic. n.d. Genuine Progress Index for Atlantic Canada http://www.gpiatlantic.org/clippings/wasteclips.htm.
Hogland, W., Kaczala, F., Jani, Y., Hogland, M., Bhatnagar, A., 2019. Beyond the zero waste concept. Linnaeus Eco-Tech. https://doi.org/10.15626/eco-tech.2014.028.
Home and Waste? 2019. What Is Zero Waste Certification - How to Create a Zero Waste Business. Waste Control Incorporated.
Home and Waste? 2020. What Is Zero Waste Certification - How to Create a Zero Waste Business. [online]. Waste Control Incorporated. Available at: https://wastecontrolinc.com/2019/05/04/what-is-zero-waste-certification-how-to-create-zero-waste/. *.
I Heard That Your Company Is Zero Waste Certified. What Does That Mean? Follow Your Heart n.d.
Intertek.com, 2020. Zero Waste to Landfill – Sustainable Landfill Diversion Solutions. [online] Available at. https://www.intertek.com/assurance/zero-waste-to-landfill/.
Markham, Ontario, Canada. Contact Erin Shapiro, eshapero@markam.ca n.d.
Meylan, G., Spoerri, A., 2014. Eco-efficiency assessment of options for metal recovery from incineration residues: a conceptual framework. Waste Manag. 34 (1), 93–100. https://doi.org/10.1016/j.wasman.2013.10.001.
New Zealand, see Zero Waste International Alliance web site. www.ZWIA.org. *.
Northeast Recycling Council, 2020b. Certified Zero Waste. Available at: https://nerc.org/news-and-updates/blog/nerc-blog/2015/10/27/certified-zero-waste. *.
Northeast Recycling Council, 2020c. Certified Zero Waste. [online] Available at. https://nerc.org/news-and-updates/blog/nerc-blog/2015/10/27/certified-zero-waste. *.
Nova Scotia program, for a governmental perspective contact Barry Friessen, barry.friessen@regional.niagara.on.ca n.d.-a.
Nova Scotia program, for a citizen's perspective contact David Wimberly, davidwimberly@eastlink.ca n.d.-b.

Novara, Italy. n.d. Contact Fabio Tomei, fabiotomei@libero.it.
Pietzsch, N., 2016. Benefits, Challenges and Critical Factors of Success for Zero Waste: A Systematic Literature Review.
San Francisco Zero waste program see videotape: "On the Road to Zero Waste, part 4: San Francicso, produced by Paul Connett for GG video, 2004, 29 minutes, GG Video, 82 Judson Street, Canton, NY 13617, contact pconnett@gmail.com and see SFenvironment web page n.d.
Song, Q., Li, J., Zeng, X., 2015. Minimizing the increasing solid waste through zero waste strategy. J. Clean. Prod. 104, 199–210. https://doi.org/10.1016/j.jclepro.2014.08.027.
Sound Resources Management Group, Inc, March 1992. Review of Waste Management Options. prepared for the City of Halifax,.
The Emerald Review, 2020. The Benefit of Zero-Waste Certification Programs. [online] Available at. http://emeraldreview.com/the-benefit-of-zero-waste-certification-programs/.
Treviso region, Italy. n.d. Contact Paolo Conto, consorzio@priula.it.
US EPA, 2020. Recycling Economic Information (REI) Report. US EPA.
Videotape: "On the Road to Zero Waste, part 1: Nova Scotia" produced by Paul Connett for GG video, 2001, 29 minutes, GG Video, 82 Judson Street, Canton, NY 13617, contact pconnett@gmail.com n.d.
Villa d'Asti, Piedmont, Italy. n.d. Contact Roberto Cavallo, http://www.youtube.com/watch?v=FpFL0e825oU&feature=related.
Waste diversion in California. Contact Rick Anthony, RicAnthony@aol.com n.d.
Zero Waste International Alliance, 2020. Zero Waste Hierarchy of Highest and Best Use - Zero Waste International Alliance*.
Zero Waste International Alliance. n.d. California counties and communities which have declared a zero waste strategy, see Zero Waste International Alliance web site: www.ZWIA.org.
Zero Waste Manifesto, 2018. Retrieved from https://www.dragon1.com/downloads/dragon1-zero-waste-manifesto.pdf.
Zerowaste Design, 2020. Available at. https://www.zerowastedesign.org/wpcontent/uploads/2017/10/ZeroWasteDesignGuidelines2017_Web.pdf. *.

Note: Sign * indicates important references.

CHAPTER THREE

Zero waste manufacturing

Sunpreet Singh[a] and Chaudhery Mustansar Hussain[b]
[a]Mechanical Engineering, National University of Singapore, Singapore, Singapore
[b]Department of Chemistry & Environmental Science, New Jersey Institute of Technology, Newark, NJ, United States

Contents

1 Introduction 45
2 Prospective of ZWM 48
  2.1 Redesigning 48
  2.2 Recycling 50
  2.3 Reuse 51
  2.4 Recovery 52
  2.5 Remanufacturing 52
3 Innovative industrial implications: Studies with ZWM view-points 54
  3.1 Zero waste of municipal incinerator fly ash (Yang et al., 2017) 54
  3.2 Three-dimensional printing for waste recycling (Somakos et al., 2016; Hunt et al., 2015; Kreiger et al., 2014; Shi et al., 2017; Pringle et al., 2018) 55
  3.3 Producing minerals from agriculturally produced waste (Sigurnjak et al., 2019) 56
  3.4 Recycling of food waste: Finding the best solution (Zhang et al., 2019) 56
  3.5 Plastic waste into liquid fuel 57
  3.6 Applications of novel methods in machining 58
  3.7 Waste for construction 60
4 Conclusions 61
References 62
Further reading 67

1 Introduction

According to the report of "World Bank," the world produced an average of about 2 billion metric tonnes of municipal solid waste in 2018 (World Bank, n.d.). It has been estimated that without implementing sustainable and green manufacturing policies, with an ultimate aim to achieve zero waste manufacturing (ZWM), it is very much possible that the quantity of the waste generation can hit 3.40 billion metric tonnes by 2050. The

Concepts of Advanced Zero Waste Tools
https://doi.org/10.1016/B978-0-12-822183-9.00003-9

report duly estimates that between one-third and 40% of waste generated worldwide is not managed properly and instead dumped or openly burned. As mentioned the United States is the biggest generator of waste per capita, followed by Denmark, and then New Zealand (Sensoneo, n.d.). Manufacturing waste, also referred as the commercial waste, hold a significant stand in the pyramid structure of solid waste. Further the manufacturing waste is the class that can utilize the maximum attention being part of the processes which are sustaining the economic growth, ranging from 15% to 25% of national economies. Acknowledging the key importance of the manufacturing sector, different governments have inaugurated the funded programs to innovate and build factories of the future.

In this today's digital market era, manufacturing of goods and services has expanded drastically, to exceptionally advanced electronic accessories, automobiles, processing equipment, synthetic tools, and even eatables, to meet up with the global demand (Lopez, 1994; Crocker, 2013). Admittedly the production of all goods and services involves a highly complex system in which resources and energy is consumed enormously (Bryson et al., 2004). These sacrificed resources, in general, are unable to recover or to reuse hence contributes toward manufacturing waste (Singh et al., 2017a). Further the manufacturing waste is not specific to a particular category of process or material, but it is common for all, resulting in the critical problem of waste for the planet (Singh et al., 2017b; Sun et al., 2017). The limited waste solutions often leaves decision-makers with no other option but to choose inefficient and environmentally polluting waste management solutions especially landfill (Cheng et al., 2019).

It has been outlined by the experts that zero waste is a whole-system approach that aims to eliminate rather than "manage" waste (Zaman, 2014), that is very difficult to meet but can be possible with repeated innovations and environmentally oriented manufacturing plans. Further the zero waste concepts represent a shift from the traditional industrial model, in which wastes are considered the norm, to integrate systems in which everything has its use in one or another way (Curran and Williams, 2012; Dey and Bhattacharyya, 2007; Elgizawy et al., 2016). The ultimate aim is to lead an industrial transformation that must have the capability to minimize the impact of industrial evolutions on the natural resources. This concept could be further classified into subdomains, including zero waste in manufacturing, zero waste of natural resources, zero emissions, maximizing the product life, zero use of the toxics, minimum waste of water, recycling, reusability, etc. (Veleva et al., 2017; Pietzsch et al., 2017; Kerdlap et al., 2019).

Many countries follow the traditional management of manufacturing waste, through landfilling and dumping in open spaces and water reservoirs, that is unhygienic and potentially disastrous to our environment. The growing manufacturing waste has made it difficult for the administrative organizations to allocate suitable landfill at reasonable costs (Cole et al., 2014). Undoubtedly, manufacturing practices should be prompted to produce multiutility end products by which it may certainly commensurate with ZWM. For this, significant attention must be devoted to formulate cutting-edge manufacturing planning focusing near-net shape manufacturing, sustainability of the products, redesigning of the products, environmental impact related to the manufacturing, postservices offered by a utilized product, ease of recycling, and cost associated with recycling (Ngoc and Schnitzer, 2009; Kumar et al., 2005). To tackle such barriers, various countries have started joint research projects within their academic institutions and industries. Singapore's Research, Innovation and Enterprise plan known as RIE2020 is one of the example of major thrust plans in Advanced Manufacturing and Engineering. Other examples from Asia include Make in India, China Manufacturing 2025, Digital Korea, and Japan's Industrial Value Chain Initiative. Germany's Factories of the Future and UK's Innovative Manufacturing are examples from Europe.

Till now the ZWM is merely a philosophical term that encourages the manufacturing systems to produce parts/needs without contributing toward waste. One such example is given by the *kerbside recycling,* Australia that produced a net worth of $72 million by recycling approximately 400,000 t of kerbside materials (Clay et al., 2007). On the same footsteps, America launched US Advanced Manufacturing Imitative, nominating the importance of digital manufacturing, sustainable manufacturing, and ZWM. Basically the main focus of such initiatives is to optimize the manufacturing processes and produce customized products for the local markets, by improving the productivity, resources efficiency, and customized mass production. On a wider scope, these initiatives encourage to eliminate all the potential hazards and unfavorable practices while manufacturing.

The prime objective of this chapter is to discuss the key prospective of ZWM in different commercial enterprises. Some special highlights of the most innovative methodologies have also been discussed for their attained quantitative and/or qualitative milestones. Further, comprehensive discussions of the recently developed sustainable and lean manufacturing systems have been made to outline their associated ZWM principles. The organization of this chapter is as follows: Section 1 introduces the background and the context of ZWM. Section 2 presents the view-points of literature on the different aspects of ZWM. Section 3 lists the innovative case studies focusing ZWM aspects. Finally, Section 4 presents the concluding remarks.

2 Prospective of ZWM

There are many options to tackle the manufacturing waste, as illustrated in Fig. 1. Depending on the type of manufacturing industry, their goals, available facilities and resources, and their prospective toward ZWM principle may differ; however, the ultimate effort will always aim to contribute toward the sustainability, including 5R's, of the goods and services as well as the environment. This section discusses the most valuable aspects of manufacturing, which is ultimately helpful in achieving ZWM principles.

2.1 Redesigning

The first R is *Redesigning*; a concept of sustainability that provides relevant information and guidelines to minimize the impact of materials and material transformations in the design process (Vezzoli and Manzini, 2008). This concept, leading toward ZWM, is often misunderstood with the conventional designing prior to manufacturing. The principle of redesigning,

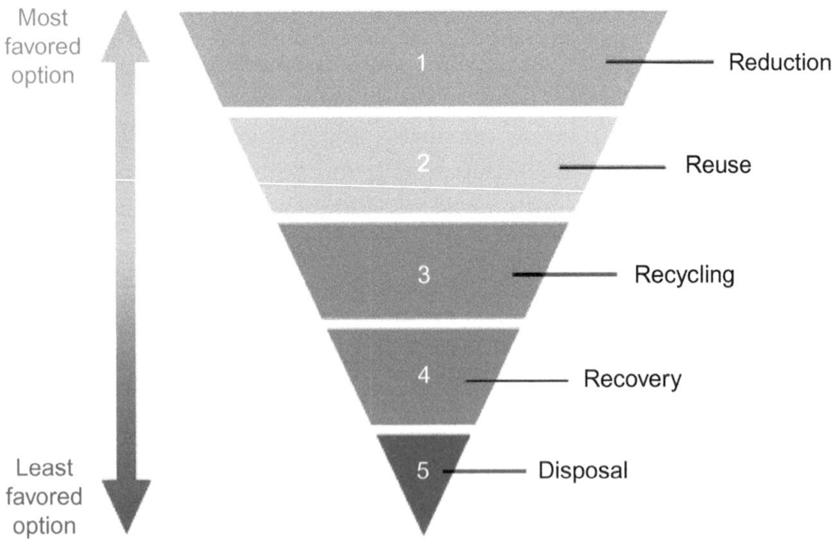

Fig. 1 Hierarchy of manufacturing waste.

aiming ZWM mainly emphasizes: (1) to achieve an optimum design to limit the appropriate use of process material and resources without compromising the quality and performance standards of the end service and (2) on repeated redesigning efforts made on a product in service for making it better in the next launch. Indeed, waste occurs at every stage of the artifact's life, during design, preconsumer, and postconsumer, and abundant wastes are usually easily available and cheap. The redesigning concept of ZWM is a creative task utilizing the consumed product to redesign for suitable services with the aid of postprocessing treatments. In the current debate on redesign the need to move from a linear economic model to a circular model is becoming increasingly evident (Helga et al., 2019). Optimizing the design, with a reliable supply chain system, can make it financially viable and harvest environmental benefits, if certain set design principles and challenges are met (Rentizelas et al., 2018). Environmentally friendly design mainly arises as a response to the need to introduce environmental criteria in the stages of production, distribution, use, recycling, and final treatment of the product with the purpose of preventing or reducing the environmental impact (Pericot et al., 2017). The environmental variables considered in addition to other conventions, such as cost, safety, and utility, should not affect the rest of the product properties. As an example to this, a US concept closely resembles waste prevention as it involves designing of materials/products to reduce their amount of toxicity before they enter the municipal solid waste stream (World Wildlife Fund and the Conservation Foundation, 1991). It is always important that the principle of designing/redesigning should focus on the efficient allocation of the product-related material flows, therefore, catching important aspects of product design and product life-cycle analysis (Eichner and Pethig, 2001). One of such studies is presented in (Fullerton and Wu, 1998), introducing a product-design variable into their model which increases the cost of producing the consumption well and stimulates recycling at the same time.

Presently the world faces two major crises: (1) financial/economical—also related to social injustice and (2) climatic—it seems that they are connected. Current practices are hopeful for the better regulations and control mechanisms for economic institutions and working toward a more just distribution of wealth; however, once the environment is destabilized, no regulations will save us from the dire consequences that scientists are predicting (Tischner and Hora, 2019). Therefore it is important to consider the environmental friendliness of the product designing process along with the performance.

2.2 Recycling

The recycling of manufacturing waste also welfares in reducing the budget energy, increase investment attractiveness, eliminate the hazard to community health, develop new occupation, lessen pollution, and progress the aesthetic condition of the territories. The second R is *Recycling,* an interrelated term to sustainability, with recycling viewed as a key issue in sustainability (Fuller et al., 1996) and dominantly as proenvironmental consumer behavior (Welfens et al., 2016). It can be clearly comprehended that waste can be significantly controlled or treated through recycling. However, recycling is often confused with reuse, which actually means the use of the waste product as process-aid, not for a new or supplementary product. Whereas recycling involves reprocessing, may be thermally, chemically, or mechanically, of the materials (not the system) through material recovery and synthesis operations. Recycling, among the others, is a behavior that may offer one fruitful pathway to a more sustainable consumer society (Schill and Shaw, 2016). As per Brundtland's Report, sustainability is defined as a development that meets the requirements of the present without compromising the ability of the future generations to meet their own needs (Brundtland, 1999).

Recycling involves defined conversion of certain types of waste into useful resources by breaking down the objects into their constituent parts (Brosius et al., 2013). Issues (Dilling, 2007) such as uncertainty of environmental science, remoteness of environmental impacts, and time lags can mean that the known impact of sustainable and recycling behaviors remain distant (Latif et al., 2018). The development of recycling allows increase in the volume of secondary raw materials used in the territory, and reduces the costs of burning and disposal of production and consumption waste (Rubinskaya et al., 2018). Even after a lot of technological developments, only a minority practice focuses on recycling, therefore, it is important to ask why recycling practices and green manufacturing continue to generate political and ethical controversies (Durant and Lucas, 2018). The answer of such question must address who is being mobilized into such reuse efforts. Moreover, it is equally important to ensure the roles to be played by the publics for future industrial-consumer activities. Although manufacturing facilities often have the greatest potential for recycling opportunities and good paybacks, they are also prone to delaying or ignoring the benefits. However, the trend for implementation of recycling programs at manufacturing facilities is definitely on the rise. There are many ways for encouraging recycling at manufacturing plants regardless of whether the facility is big or small (Kennedy, n.d.).

2.3 Reuse

The 3rd R, reuse of the manufacturing waste, means to use the discarded products without any transformation and without changing the shape or original nature (Hernández-Apaolaza et al., 2005). This is an option in the waste hierarchy. Different types of manufacturing waste, including plastic, metallic burs, defective goods and edible items, stationary, books, and anything else that can be used again for a similar or different application (Ismail and Al-Hashmi, 2008). Reuse means that less manufacturing waste will be dumped, and reduce the amount of waste that has to be disposed of helps to reduce these problems (Li, 2009; Demir, 2009). Some other advantages of waste reuse are as follows: (1) community benefits—through distributing the manufacturing waste to the community, engaging in job-training programs and general training, (2) economic benefits—means reusing manufacturing waste for creating new products from raw materials, and (3) environmental benefits—reusing something without water, energy, or other resources is completely environmental friendly.

For this the generators of industrial wastes need to manage the wastes themselves or contact qualified industrial treatment companies to dispose of their industrial wastes. However, many still lack skills, manpower, economic scale, technology and equipment, and pioneering vision for industrial waste management (Wei and Huang, 2001). In Pacelli et al. (2015) a methodology has been presented aiming to enable designers to develop products based on scrap reuse. In this approach the designers evaluated whether the operation is economically and environmentally advantageous compared to the realization of the same product using new raw materials or new half-finished components, adopting standard industrial procedures. Author's methodology was constituted by three main sequential steps:

- Phase 1—Scrap optimization
- Phase 2—Unavoidable scrap analysis
- Phase 3—Designing with scraps

Similarly, different models have been presented in the literature dealing with ceramic (Pacheco-Torgal and Jalali, 2010; Penteado et al., 2016), glass (Corinaldesi et al., 2005; Olofinnade et al., 2017), metals and alloys (Davidson et al., 2009; Brunori et al., 2005), plastics (Panyakapo and Panyakapo, 2008; Goodship, 2009), agricultural (Sheets et al., 2015; Bradford et al., 2008), and other types of wastes (Jiménez and Asano, 2008; Russ and Meyer-Pittroff, 2004; Banković-Ilić et al., 2014).

2.4 Recovery

The manufacturing waste which cannot be used for aiding a secondary manufacturing process could be converted into energy, such a heat and electricity (Beegle and Borole, 2018; Khalid et al., 2018; Ross, 2018). However, recovering energy from the manufacturing waste belongs to a subsidiary part of the 3rd R, recovery. This, indeed, is the most expensive treatment of the waste materials and mainly suitable for only large amount of wastes. In case of chemical treatment, reuse of the manufacturing waste is one of the most advanced technologies used to transform waste plastic materials into smaller molecules, usually liquids or gases, which are suitable for use as a feedstock for the production of new petrochemicals and plastics (Arena and Mastellone, 1999). Chemically reuse of manufacturing waste can be categorized into advanced process such as pyrolysis, gasification, liquid-gas hydrogenation, viscosity breaking, steam or catalytic cracking, and the use of waste as a reducing agent in blast furnaces. A number of environmental concerns are associated with this method, mainly emission of certain air pollutants such as CO_2, NO_x, and SO_x.

To enhance energy conservation and emission reduction, and promote construction of energy-efficient systems, a series of measures and multiple binding targets have been identified by China and others (Ministry of Industry and Information Technology, n.d.; Zhang et al., 2017). In Ammar et al. (2012), authors addressed the potential for low-grade heat recovery with regard to new incentives and technological advances and found that the benefit of recycling and utilizing low-grade thermal energy is highly dependent on the qualities and properties of the waste streams heat. Further the authors developed a simulation model and applied exergy analysis to optimize steel production and recycling system from various view-points (Shigaki et al., 2002). Even for recovering the waste, heat from industrial processes can replace fossil fuels in various applications results in improvements of energy efficiency and in reduced carbon emissions (Vourdoubas, 2018). It has been found that the fossil fuel–fired power plants produce large quantities of waste heat, which creates heat pollution. Therefore it is essential to recover the heat from the industrial wastes (Loibl et al., 2017).

2.5 Remanufacturing

Remanufacturing, 5th R, is one of the important domain of the manufacturing industry, symbolizing sustainable development idea in manufacturing industry. It is the technical process of using waste parts, objects and original sizes, or even their service performance (Zhou et al., 2016). It is of great significance for protecting shortage of resources and also protecting the environment (Yenipazarli, 2016). It is recognized to be an effective strategy in closing the loop by enhancing resource efficiency, through reuse of components and products, as input material and guarantees competitive advantages by significant price reductions. When compared with routine product manufacturing, many special features in remanufacturing, including dispersion of eminence and foundations of objects to be manufactured and the uncertainty of the process parameters (Narayana, 2016). The concept of generalized remanufacturing is used for technology process analysis within remanufacturing, as referred to repurchase, dismantling, repair, recycling, and material utilization after the end of product utilization process (Deng et al., 2017).

After waste products are collected in the processing enterprises, they are usually disassembled and cleaned. Since the conditions of waste products are different, and their brands and specification parameters are varied, large-scale processing is difficult. After disassembling, good-quality products can be right reused, for example, traded as overhaul parts. For shares with remanufacturing value, surface engineering technologies, such as cladding and plasma cladding can be used to repair them to restore their functions, which once tested can be used as parts for new product manufacturing or be sold directly in markets (Kurilova-Palisaitiene et al., 2018; Matsumoto et al., 2016; Xiong et al., 2016).

Remanufacturing strategies can lead to sustainability in the production and consumption of electronics products by increasing the lifetime cycle of manufacturing waste (Singhal et al., 2019). In countries like India, remanufacturing is currently at a very basic state; however, the United States, Germany, and the United Kingdom are directing remanufacturing activities. Developing countries' remanufacturing practices are becoming more important, as the involvement of international enterprises, liberalization in investment and trade policies, and overconsumption of natural resources are encouraging catalysts (Chaowanapong et al., 2018). It is the need of the hour, to consider life cycle assessment, cost benefit analysis, and life cycle cost are the prominent tools to incorporate to actualize and improve the remanufacturing scheme for the sake of minimizing environmental impact and to establish the sustainability of the remanufacturing system (Zlamparet et al., 2017). For further strengthening, mixed evaluation model based on merit, limitation, avenue and hazard analysis, and analytic hierarchy process are fruitful.

3 Innovative industrial implications: Studies with ZWM view-points

The overconsumption of manufactured goods has turned the commercial manufacturing into a challenging issue as the waste associated is being considered as a global challenge (Mahdavi and Danesh, 2016). Production activities, across the globe, are posing remarkably negative impact on the environment and the effective countermeasures in the reduction of waste and energy conservation are strongly in demand. King and Lenox (King and Lenox, 2001) described "green" as 'the good public spillover of Lean" and explain these positive effects in the efforts toward waste reduction. It involves green design of products, use of environmental friendly raw materials, eco-friendly packing, distribution, and reuse after end-of-life of a product. The aim of this section is to discuss the most innovative industrial implications (III) strategies of ZWM concept in the different manufacturing sectors.

3.1 Zero waste of municipal incinerator fly ash (Yang et al., 2017)

This case study presented a practice followed by the researchers of the National Sun Yat-Sen University of Taiwan in 2017. Considering zero waste of municipal incinerator fly ash (MIFA) as their main objective, this work focused on the melting of the fly ash in electric arc furnaces (EAFs) of steel-making operation and satisfactorily dealt with the treatment of several hundred tons of incinerator fly ash as a process similar to molten salt oxidation. The designed treatment capacities for the plants 1 and 2 were 1350 and 900 metric tonnes/day, respectively, and the innovative technology used is shown in Fig. 2.

The work showed no solid evidence to show the whereabouts of waste fly ash injected and melted in steel-making EAF, however, concerning the occurrence and fate of MIFA after melting in a steel-making EAF the relevant legal considerations should be obeyed. On the other hand, by using incinerator fly ash while steelmaking, it has the ability to convert nonrecyclable and hazardous incinerator fly ash to various types of recyclable materials including dust, oxidized slag, and reduced slag. This means that material is not 100% recycled at the present time, but, at least landfilling can be avoided. Furthermore the development of new methods for increasing the recycling rate of such material has a futuristic mission and grand opportunity to work on.

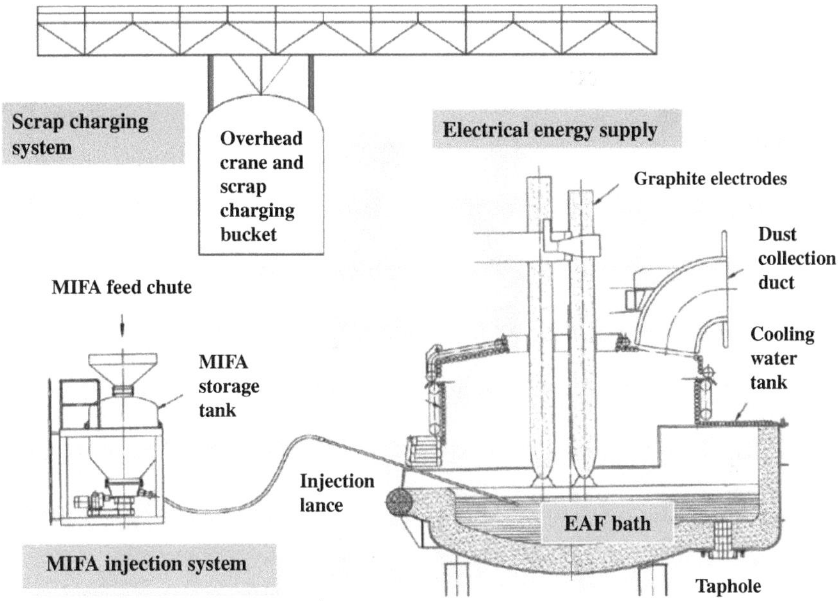

Fig. 2 Innovative melting process of municipal incinerator fly ash for steel making (Yang et al., 2017).

3.2 Three-dimensional printing for waste recycling (Somakos et al., 2016; Hunt et al., 2015; Kreiger et al., 2014; Shi et al., 2017; Pringle et al., 2018)

It has been found that the innovative three-dimensional (3D) printing platform for recycling wastes found significant potential in developing knowledge in identification of electrical and electronic devices and types and qualities of materials embodied, suitable for printing (Somakos et al., 2016). A year of pilot study for developing and testing in different participating countries has contributed a lot in this mission. The MTU (United States) (Hunt et al., 2015; Kreiger et al., 2014) has estimated the life cycle analysis of recycled high-density polythene (HDPE) printed products for their environmental aspects. The results of the study showed that distributed recycling of postconsumer HDPE for 3D printing filament uses less embodied energy than the best-case scenario investigated for a high-population density city using centralized recycling. A similar study (Shi et al., 2017) demonstrated a new 3D printing material, vitrimer epoxy, which can be recycled and turned into parts with complicated geometries as highly desirable because of the advantages offered by such material. The new ink and method developed in this work provided a unique platform of postconsumer-based thermocurable thermosetting polymer that otherwise is extremely difficult to use.

Further the four steps based on such approach have been used by the researchers to develop wood polymer composites for 3D printing of furniture wood. In this work, wood-based waste material was received from several furniture manufacturing companies, which were reduced to the workable size scale (about 80 μm). The material was mixed with the matrix polymer, in second step, and was extruding into filament of homogeneous thickness and density, in the third step. Ultimately the developed filament was used in the 3D printer to form the final 3D-printed product (Pringle et al., 2018). Many such examples are available in the literature (Griffiths et al., 2016; Mohammed et al., 2017; Tran et al., 2017; Kumar et al., 2019; Chong et al., 2017), for the waste utilization. The 3D printing platform does not only create new recycling opportunities but also has potential to control the waste associated with the manufacturing. The most appealing possibility of this digital manufacturing avenue is that the machines could recycle the waste polymers themselves and reuse them as feedstock (Pasricha and Greeninger, 2018). In addition, it has been believed that zero waste economy could be boomed by bringing the promising 3D printing industry with experts dealing with the waste management.

3.3 Producing minerals from agriculturally produced waste (Sigurnjak et al., 2019)

This study described pathways to recover ammonia, from agricultural waste, with an aim to produce ammonium nitrate (AN) and ammonium sulfate (AS), evaluate fertilizer performance of the recovered greenhouse (*Lactuca sativa* L.) and full field (*Zea mays* L.), and compare the observed performances. The results of the study highlighted that observed variability in concentrations (N) of these products is seen as the biggest challenge for their recognition as fertilizers. Both AN and AS showed similar traits as synthetic fertilizers, demonstrated in pot and field experiments. Further, AS led to a similar effect on crop yield and risk for nitrate leaching as compared to conventional synthetic fertilizer.

3.4 Recycling of food waste: Finding the best solution (Zhang et al., 2019)

This is a case study of a Chinese restaurant wherein an entire life cycle of the anaerobic digestion–aerobic composting technique of restaurant food waste recycling has been discussed, in contrast of landfill and incineration. As it is well known that the life cycle of an anaerobic digestion–aerobic composting is a prime consumer of water, clay, coal, crude oil, and natural gas, therefore preprocessing phase and anaerobic digestion results in a great environmental hazard. It has been reported that the most influential factor generated during preprocessing, solid composting, and biogas slurry composting is ammonia, whereas the most influential factor produced during collection and transportation, biogas power generation, and heating is nitrogen oxides.

Further, sulfur oxides are the most influential factor produced during anaerobic digestion. Among the three methods of restaurant food waste treatment, such as anaerobic digestion–aerobic composting, incineration, and landfill disposal, the results favored the anaerobic digestion–aerobic composting technique due to their least environmental impact. Further the landfill method had a medium impact; however, this method includes its addition costs in terms of transportation and depletion of natural views. In particular the latter two techniques may significantly exacerbate global warming.

3.5 Plastic waste into liquid fuel

Being most extensively used, plastic is accumulated in a massive quantity and either takes a long time to decompose or never decomposes. However, the plastic can be smartly handled by the manufacturing companies by either cycling it into feedstock or alternatively can be converted into fuel oil (Bow and Pujiastuti, 2019). Pyrolysis of plastic, as a chemical decomposition method, through which the waste-manufactured plastic or processed waste plastic can be broken into gas phase with limited use of heating and oxygen (Oasmaa et al., 2020; Miandad et al., 2019). In (Ding et al., 2019), catalytic microwave-assisted pyrolysis, refer Fig. 3 for schematic, of low-density polyethylene was performed to simultaneously improve yield and quality of gasoline-range products.

Based on the outcomes, following are the key benefits offered by the adopted technique:

- The reaction time and operating cost has been reduced by rapid and selective heating of microwave-assisted pyrolysis.
- The setup of catalyst bed and pyrolysis zone is fruitful in proceeding both pyrolysis and catalytic reform at their optimum conditions.
- Adding a small amount of nitric oxide (NiO) can efficiently improve the quality of the oil product, which will have a minor effect on the operating cost.

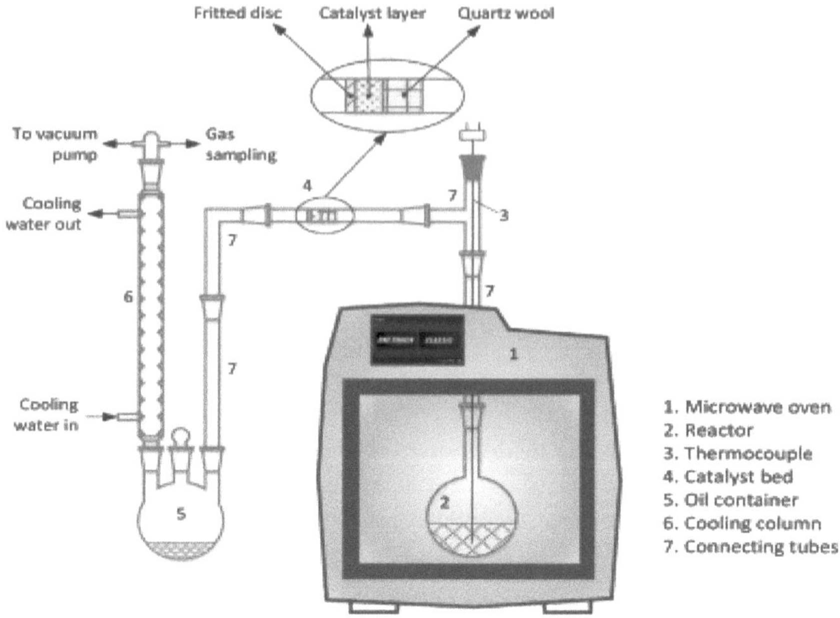

Fig. 3 Schematic of microwave-assisted pyrolysis (Ding et al., 2019).

- The system is simple and straightforward without many complicated structures; thus, it is easy to scale up.

3.6 Applications of novel methods in machining

Machining, a major manufacturing operation, involves a numerous sustainability factors, which have a huge impact on environmental due to excessive use of coolant and lubricant, and energy consumption. Further, it has been observed that the relationship between machining technologies and environmental impact remains insufficiently discussed and the environmental impact are little evaluated (Dietmair and Verl, 2009). In (Chauhan et al., 2015), authors presented the production system to identify prominent issues of wastes existing within an organization that can give a motivation for application of lean principles. A number of research efforts have been made to conceive dry or minimal quantity lubrication-based machining processes to reduce the emission of hazardous gaseous and manufacturing wastes, burrs (Pusavec et al., 2010).

It has been found that cryogenic-assisted machining is completely clean and has the highest sustainability potential. In (Pušavec and Kopač, 2011), authors performed study of sustainable cryogenic and measured the tool life to determine the total production cost per part, covering all sustainability measures. Sreejith and Ngoi (Sreejith and Ngoi, 2000) highlighted the recent advancements in the direction of dry machining and suggested the industries to use this approach for all types of processes as the supply of cooling media is through air and oil in the form of an aerosol. A detailed picture of near dry machining is shown in Fig. 4.

Nanotechnology can also help the industries to eliminate the potential environmental threat, substantially (Hussain and Mishra, 2018; Hussain and Kecili, 2019; Hussain, 2019). Authors (Khan et al., 2019) presented a well-explained case study to outline the importance of optimize cutting parameters to reduce the total specific energy consumed by the milling of AISI D2 steel. Further, dry cutting methodology has been considered to decrease the environmental impacts, through sustainability aspects, linked to the coolant usage [minimal quantity lubrication (MQL) and nanofluid-based minimal quantity lubrication (NFMQL)]. The comprehensive analysis highlighted the superiority of using NFMQL in comparison to MQL. The NFMQL resulted in the reduction of temperature, at the tool-workpiece interface, from 16.2% to 34.5% and surface roughness of the machined product from 11.3% to 12%. The reduction in the tool/workface temperature always favors the tool life as impact of thermal shocks while machining could be decreased and, ultimately, increase the tool life. Further the fall in temperature, being the prime cause of evaporation of the coolant, will reduce the formation of hazardous gases, therefore, support sustainability. Indeed,

Fig. 4 Example of near dry machining (Singh et al., 2017a).

as discussed in the study, the use of NFMQL complemented the innovation essential for sustainable production by better machinability and integrated applications.

3.7 Waste for construction

The natural ways of reusing inorganic industrial wastes is to use them for the production of building materials, specifically raw materials in the concrete manufacture. This is the only way of recycling that has positive impact on the environment, by reducing the amount of waste and limits mining of additional mineral aggregate deposits, inorganic ceramic waste needs no special processing when used as an aggregate, producing the concrete mix with aggregate using recycled ceramic ware (Halicka et al., 2013).

The recent experiments on the use of waste ceramics have been primarily focused on its environmental impact. The crushed ceramic waste is used as a mixture to traditional collections or as a substitute for some part of aggregate of a chosen dimension (Bektas et al., 2009), and the powder obtained from crushed ceramic waste can be used as pozzolanic admixture to Portland cement (Binici et al., 2012). Apart from waste ceramics, plastics can also be recycled to produce new materials, cement composites that appears as one of the best solution for disposing of plastic. This applies to all classes of polymers, including polyethylene terephthalate bottle (Yesilata et al., 2009), polyvinyl chloride pipe (Kou et al., 2009), HDPE (Naik et al., 1996), thermosetting plastics (Panyakapo and Panyakapo, 2008), shredded waste (Al-Manaseer and Dalal, 1997), expandable polystyrene (Kan and Demirboğa, 2009), glass-reinforced plastic (Asokan et al., 2010), polycarbonate (Hannawi et al., 2010), polyurethane (Fraj et al., 2010), and polypropylene (PP) (Bayasi and Zeng, 1993). Moreover, information appeared in several publications provided a clearer picture on the characteristics of concrete containing reinforced plastics (Bayasi and Zeng, 1993; Siddique et al., 2008).

The case study by Giri et al. (Giri et al., 2020) attempted to explore the implication of waste recycled concrete aggregates and waste milk packaging polyethylene in bituminous concrete mixes. The dense bituminous macadam mixtures were prepared with coarse recycled aggregates and cement and stone dust as filler. The prepared concrete mixtures were tested for their tensile strength, retained stability, dynamic modulus, and wheel tracking rutting. It has been found that recycled concrete aggregates mixed with waste polyethylene and cement as filler showed higher stability, may be due to the rougher surface recycled concrete aggregates. Further, maximum indirect tensile strength has been observed for mixture prepared with natural aggregates, cement, and waste polyethylene when compared to the others. Higher dynamic modulus has been found with natural aggregates. Generally the strength characteristics of bituminous mixes with waste milk polypropylene are better at higher temperature. Further, wheel tracking tests resulted in minimum rut depth for natural aggregates, cement with waste polyethylene.

4 Conclusions

The literature review on ZWM has been seen as the futuristic research wave of the industrial revolution. Considering cues from the green, lean, or sustainable manufacturing, it is the only methodology of boosting industrial economy that can seek either to eliminate nonproductive output or transform the same into product output that generates values, directly or indirectly. Increasingly, manufacturers are availing the opportunities to consider industrial wastes as subsidiary materials which can aid supplementary manufacturing processed. Implementing various zero waste initiatives enables organizations to lower production costs, strengthen bottom lines, environmental sustainability, and reduction in the consumption of natural resources.

Although waste elimination is a big challenge for existing manufacturing world, it is a common belief that manufacturing rubbish can be significantly controlled through adopting the five prospective of ZWM. Indeed, literature advocates that reuse or recycle of industrial wastes need matured regulations and attitude. It is the most important requirement that the government should bring environmental friendly manufacturing policies into action and this demands funded start-ups to establish public waste treatment plants to recover materials in the form of feedstock, heat, or energy. Subsequently, it will help to drop the landfilling practices with a huge impact. Further, established design for manufacturing, design of experimentation and simulation software are also important at the premanufacturing stages to minimize the use of materials, energy, and resources.

There are multiple techniques that manufacturers can apply to their operations to minimize their waste streams. Many such examples included have been comprehensively discussed in the literature review are discussed in this chapter.

References

Al-Manaseer, A.A., Dalal, T.R., 1997. Concrete containing plastic aggregates. Concr. Int. 19 (8), 47–52.

Ammar, Y., Joyce, S., Norman, R., Wang, Y., Roskilly, A.P., 2012. Low grade thermal energy sources and uses from the process industry in the UK. Appl. Energy 89 (1), 3–20.

Arena, U., Mastellone, M.L., 1999. Particle Agglomeration during Energy Recovery from Plastic Wastes by Means of Fluidized Bed Reactors. Univ. Federico II of Naples, Napoli (IT).

Asokan, P., Osmani, M., Price, A.D., 2010. Improvement of the mechanical properties of glass fibre reinforced plastic waste powder filled concrete. Constr. Build. Mater. 24 (4), 448–460.

Banković-Ilić, I.B., Stojković, I.J., Stamenković, O.S., Veljkovic, V.B., Hung, Y.T., 2014. Waste animal fats as feedstocks for biodiesel production. Renew. Sust. Energ. Rev. 32, 238–254.

Bayasi, Z., Zeng, J., 1993. Properties of polypropylene fiber reinforced concrete. Mater. J. 90 (6), 605–610.

Beegle, J.R., Borole, A.P., 2018. Energy production from waste: evaluation of anaerobic digestion and bioelectrochemical systems based on energy efficiency and economic factors. Renew. Sust. Energ. Rev. 96, 343–351.

Bektas, F., Wang, K., Ceylan, H., 2009. Effects of crushed clay brick aggregate on mortar durability. Constr. Build. Mater. 23 (5), 1909–1914.

Binici, H., Kapur, S., Arocena, J., Kaplan, H., 2012. The sulphate resistance of cements containing red brick dust and ground basaltic pumice with sub-microscopic evidence of intra-pore gypsum and ettringite as strengtheners. Cem. Concr. Compos. 34 (2), 279–287.

Bow, Y., Pujiastuti, L.S., 2019. Pyrolysis of polypropylene plastic waste into liquid fuel. In: IOP Conference Series: Earth and Environmental Science. Vol. 347, No. 1. IOP Publishing, p. 012128.

Bradford, S.A., Segal, E., Zheng, W., Wang, Q., Hutchins, S.R., 2008. Reuse of concentrated animal feeding operation wastewater on agricultural lands. J. Environ. Qual. 37 (5_Supplement), S-97.

Brosius, N., Fernandez, K.V., Cherrier, H., 2013. Reacquiring consumer waste: treasure in our trash? J. Public Policy Mark. 32 (2), 286–301.

Brundtland, G.H., 1999. Environment and health: Europe's partnerships can be a model. BMJ 318 (7199), 1635–1636. https://doi.org/10.1136/bmj.318.7199.1635.

Brunori, C., Cremisini, C., Massanisso, P., Pinto, V., Torricelli, L., 2005. Reuse of a treated red mud bauxite waste: studies on environmental compatibility. J. Hazard. Mater. 117 (1), 55–63.

Bryson, J.R., Daniels, P.W., Warf, B., 2004. Service Worlds: People, Organisations and Technologies. Psychology Press.

Chaowanapong, J., Jongwanich, J., Ijomah, W., 2018. The determinants of remanufacturing practices in developing countries: evidence from Thai industries. J. Clean. Prod. 170, 369–378.

Chauhan, P., Rangrej, S., Samvatsar, K., Sheth, S., 2015. Application of lean manufacturing principles for process time reduction–a case of conveyor pulley manufacturing. In: Proceedings of 5th National Conference on Recent Advances in Manufacturing (RAM-2015), pp. 423–428.

Cheng, Z., Sun, Z., Zhu, S., Lou, Z., Zhu, N., Feng, L., 2019. The identification and health risk assessment of odor emissions from waste landfilling and composting. Sci. Total Environ. 649, 1038–1044.

Chong, S., Pan, G.T., Khalid, M., Yang, T.C., Hung, S.T., Huang, C.M., 2017. Physical characterization and pre-assessment of recycled high-density polyethylene as 3D printing material. J. Polym. Environ. 25 (2), 136–145.
Clay, S., Gibson, D., Ward, J., 2007. Sustainability Victoria: influencing resource use, towards zero waste and sustainable production and consumption. J. Clean. Prod. 15 (8–9), 782–786.
Cole, C., Osmani, M., Quddus, M., Wheatley, A., Kay, K., 2014. Towards a zero waste strategy for an English local authority. Resour. Conserv. Recycl. 89, 64–75.
Corinaldesi, V., Gnappi, G., Moriconi, G., Montenero, A., 2005. Reuse of ground waste glass as aggregate for mortars. Waste Manag. 25 (2), 197–201.
Crocker, R., 2013. From access to excess: consumerism, 'compulsory' consumption and behaviour change. In: Motivating Change: Sustainable Design and Behaviour in the Built Environment. Routledge, pp. 43–64.
Curran, T., Williams, I.D., 2012. A zero waste vision for industrial networks in Europe. J. Hazard. Mater. 207, 3–7.
Davidson, J., Good, C., Welsh, C., Brazil, B., Summerfelt, S., 2009. Heavy metal and waste metabolite accumulation and their potential effect on rainbow trout performance in a replicated water reuse system operated at low or high system flushing rates. Aquac. Eng. 41 (2), 136–145.
Demir, I., 2009. Reuse of waste glass in building brick production. Waste Manag. Res. 27 (6), 572–577.
Deng, C., Liu, Y., Zhao, J., Wei, B., Yin, G., 2017. Analysis of the machine tool dynamic characteristics in manufacturing space based on the generalized dynamic response model. Int. J. Adv. Manuf. Technol. 92 (1–4), 1411–1424.
Dey, S., Bhattacharyya, K.K., 2007. Split and collectorless flotation to medium coking coal fines for multi-product zero waste concept. Fuel Process. Technol. 88 (6), 585–590.
Dietmair, A., Verl, A., 2009. A generic energy consumption model for decision making and energy efficiency optimisation in manufacturing. Int. J. Sustain. Eng. 2 (2), 123–133.
Dilling, L., 2007. The opportunities and responsibility for carbon cycle science in the US. Environ. Sci. Pol. 10 (1), 1–4.
Ding, K., Liu, S., Huang, Y., Liu, S., Zhou, N., Peng, P., Wang, Y., Chen, P., Ruan, R., 2019. Catalytic microwave-assisted pyrolysis of plastic waste over NiO and HY for gasoline-range hydrocarbons production. Energy Convers. Manag. 196, 1316–1325.
Durant, D., Lucas, A., 2018. Manufacturing a better planet: challenges arising from the gap between the best intentions and social realities. Theatr. Rec. 3 (2), 17.
Eichner, T., Pethig, R., 2001. Product design and efficient management of recycling and waste treatment. J. Environ. Econ. Manag. 41 (1), 109–134.
Elgizawy, S.M., El-Haggar, S.M., Nassar, K., 2016. Approaching sustainability of construction and demolition waste using zero waste concept. Low Carbon Econ. 7 (1), 1.
Fraj, A.B., Kismi, M., Mounanga, P., 2010. Valorization of coarse rigid polyurethane foam waste in lightweight aggregate concrete. Constr. Build. Mater. 24 (6), 1069–1077.
Fuller, D.A., Allen, J., Glaser, M., 1996. Materials recycling and reverse channel networks: the public policy challenge. J. Macromark. 16 (1), 52–72.
Fullerton, D., Wu, W., 1998. Policies for green design. J. Environ. Econ. Manag. 36 (2), 131–148.
Giri, J.P., Panda, M., Sahoo, U.C., 2020. Use of waste polyethylene for modification of bituminous paving mixes containing recycled concrete aggregates. Road Mater. Pav. Des. 21 (2), 289–309.
Goodship, V. (Ed.), 2009. Management, Recycling and Reuse of Waste Composites. Elsevier.

Griffiths, C.A., Howarth, J., De Almeida-Rowbotham, G., Rees, A., Kerton, R., 2016. A design of experiments approach for the optimisation of energy and waste during the production of parts manufactured by 3D printing. J. Clean. Prod. 139, 74–85.
Halicka, A., Ogrodnik, P., Zegardlo, B., 2013. Using ceramic sanitary ware waste as concrete aggregate. Constr. Build. Mater. 48, 295–305.
Hannawi, K., Kamali-Bernard, S., Prince, W., 2010. Physical and mechanical properties of mortars containing PET and PC waste aggregates. Waste Manag. 30 (11), 2312–2320.
Helga, A., Rognoli, V., Carla, L., 2019. Re-designing recovered materials. Case study: fiberglass in the nautical sector. In: The 3rd LeNS World Distributed Conference. Poli. Design, pp. 884–889.
Hernández-Apaolaza, L., Gascó, A.M., Gascó, J.M., Guerrero, F., 2005. Reuse of waste materials as growing media for ornamental plants. Bioresour. Technol. 96 (1), 125–131.
Hunt, E.J., Zhang, C., Anzalone, N., Pearce, J.M., 2015. Polymer recycling codes for distributed manufacturing with 3-D printers. Resour. Conserv. Recycl. 97, 24–30.
Hussain, C.M., 2019. Handbook of Environmental Materials Management, first ed. Springer International Publishing.
Hussain, C.M., Kecili, R., 2019. Modern Environmental Analysis Techniques for Pollutants. Elsevier.
Hussain, C.M., Mishra, A.K., 2018. Nanotechnology in Environmental Science, 2 Volumes. John Wiley & Sons.
Ismail, Z.Z., Al-Hashmi, E.A., 2008. Reuse of waste iron as a partial replacement of sand in concrete. Waste Manag. 28 (11), 2048–2053.
Jiménez, B., Asano, T. (Eds.), 2008. Water Reuse: An International Survey of Current Practice, Issues and Needs. IWA, London.
Kan, A., Demirboğa, R., 2009. A novel material for lightweight concrete production. Cem. Concr. Compos. 31 (7), 489–495.
Kennedy, M., n.d. Wastecare. https://www.wastecare.com/Articles/Waste_Reduction_Recycling_Tips_Manufacturing.htm.
Kerdlap, P., Low, J.S., Ramakrishna, S., 2019. Zero waste manufacturing: a framework and review of technology, research, and implementation barriers for enabling a circular economy transition in Singapore. Resour. Conserv. Recycl. 151, 104438.
Khalid, M., Johnson, E., Vij, A., Bouwer, E.J., Janjua, H.A., Betenbaugh, M.J., 2018. Anaerobic digestion restricted to phase I for nutrient release and energy production using waste-water grown Chlorella vulgaris. Chem. Eng. J. 352, 756–764.
Khan, A.M., Jamil, M., Haq, A.U., Hussain, S., Meng, L., He, N., 2019. Sustainable machining. Modeling and optimization of temperature and surface roughness in the milling of AISI D2 steel. Ind. Lubr. Tribol. 71 (2), 267–277.
King, A.A., Lenox, M.J., 2001. Lean and green? An empirical examination of the relationship between lean production and environmental performance. Prod. Oper. Manag. 10 (3), 244–256.
Kou, S.C., Lee, G., Poon, C.S., Lai, W.L., 2009. Properties of lightweight aggregate concrete prepared with PVC granules derived from scraped PVC pipes. Waste Manag. 29 (2), 621–628.
Kreiger, M.A., Mulder, M.L., Glover, A.G., Pearce, J.M., 2014. Life cycle analysis of distributed recycling of post-consumer high density polyethylene for 3-D printing filament. J. Clean. Prod. 70, 90–96.
Kumar, V., Bee, D.J., Shirodkar, P.S., Tumkor, S., Bettig, B.P., Sutherland, J.W., 2005. Towards sustainable "product and material flow" cycles: Identifying barriers to achieving product multi-use and zero waste. In: Proceedings of International Mechanical Engineering Congress and Exposition, November 5–11, Orlando, FL.
Kumar, S., Singh, R., Singh, T.P., Batish, A., 2019. On investigation of rheological, mechanical and morphological characteristics of waste polymer-based feedstock filament for 3D printing applications. J. Thermoplast. Compos. Mater. 0892705719856063.

Kurilova-Palisaitiene, J., Sundin, E., Poksinska, B., 2018. Remanufacturing challenges and possible lean improvements. J. Clean. Prod. 172, 3225–3236.
Latif, S.A., Omar, M.S., Bidin, Y.H., Awang, Z., 2018. Analyzing the effect of situational factor on recycling behaviour in determining the quality of life. J. Asian Behav. Stud. 3 (6), 11–17.
Li, X., 2009. Recycling and reuse of waste concrete in China: part II. Structural behaviour of recycled aggregate concrete and engineering applications. Resour. Conserv. Recycl. 53 (3), 107–112.
Loibl, W., Stollnberger, R., Österreicher, D., 2017. Residential heat supply by waste-heat re-use: sources, supply potential and demand coverage—a case study. Sustainability 9 (2), 250.
Lopez, R., 1994. The environment as a factor of production: the effects of economic growth and trade liberalization. J. Environ. Econ. Manag. 27 (2), 163–184.
Mahdavi, E., Danesh, S.S., 2016. How to reduce brass components wastage and improve product quality using design of experiments (DOE) techniques (Case Application at Fardan electric Co)? Int. J. Human. Cult. Stud. 2356-5926., 1919–1936.
Matsumoto, M., Yang, S., Martinsen, K., Kainuma, Y., 2016. Trends and research challenges in remanufacturing. Int. J. Precis. Eng. Manuf. - Green Technol. 3 (1), 129–142.
Miandad, R., Rehan, M., Barakat, M.A., Aburiazaiza, A.S., Khan, H., Ismail, I.M., Dhavamani, J., Gardy, J., Hassanpour, A., Nizami, A.S., 2019. Catalytic pyrolysis of plastic waste: moving towards pyrolysis based biorefineries. Front. Energy Res. 7, 27.
Ministry of Industry and Information Technology. n.d. Guidance on energy conservation in iron and steel industry http://www.miit.gov.cn/n1146295/n1146592/n1146754/n1235000/n1235002/n1235003/n1235005/c3096412/content.html.
Mohammed, M., Mohan, M., Das, A., Johnson, M.D., Badwal, P.S., McLean, D., Gibson, I., 2017. A low carbon footprint approach to the reconstitution of plastics into 3D-printer filament for enhanced waste reduction. In: DesTech 2016: Proceedings of the International Conference on Design and Technology. Knowledge E, pp. 234–241.
Naik, T.R., Singh, S.S., Huber, C.O., Brodersen, B.S., 1996. Use of post-consumer waste plastics in cement-based composites. Cem. Concr. Res. 26 (10), 1489–1492.
Narayana, S., 2016. Uncertainty analysis in product service system: Bayesian network modelling for availability contract (Doctoral dissertation). Loughborough University.
Ngoc, U.N., Schnitzer, H., 2009. Sustainable solutions for solid waste management in southeast Asian countries. Waste Manag. 29 (6), 1982–1995.
Oasmaa, A., Qureshi, M.S., Pihkola, H., Deviatkin, I., Mannila, J., Tenhunen, A., Minkkinen, H., Pohjakallio, M., Laine-Ylijoki, J., 2020. Pyrolysis of plastic waste: opportunities and challenges. J. Anal. Appl. Pyrolysis 104804.
Olofinnade, O.M., Ede, A.N., Ndambuki, J.M., 2017. Sustainable green environment through utilization of waste soda-lime glass for production of concrete. J. Mater. Environ. Sci. 8 (4), 1139–1152.
Pacelli, F., Ostuzzi, F., Levi, M., 2015. Reducing and reusing industrial scraps: a proposed method for industrial designers. J. Clean. Prod. 86, 78–87.
Pacheco-Torgal, F., Jalali, S., 2010. Reusing ceramic wastes in concrete. Constr. Build. Mater. 24 (5), 832–838.
Panyakapo, P., Panyakapo, M., 2008. Reuse of thermosetting plastic waste for lightweight concrete. Waste Manag. 28 (9), 1581–1588.
Pasricha, A., Greeninger, R., 2018. Exploration of 3D printing to create zero-waste sustainable fashion notions and jewelry. Fash. Text. 5 (1), 1–8.
Penteado, C.S., de Carvalho, E.V., Lintz, R.C., 2016. Reusing ceramic tile polishing waste in paving block manufacturing. J. Clean. Prod. 112, 514–520.
Pericot, N.G., Sierra, E.M., del Río, M.M., Carrasco, Ó.L., 2017. Packaging waste reduction in construction sites: eco-redesign of cardboard boxes. Waste Recycl. Res. 1. https://doi.org/10.4081/wrr.2017.6.

Pietzsch, N., Ribeiro, J.L., de Medeiros, J.F., 2017. Benefits, challenges and critical factors of success for zero waste: a systematic literature review. Waste Manag. 67, 324–353.
Pringle, A.M., Rudnicki, M., Pearce, J.M., 2018. Wood furniture waste–based recycled 3-D printing filament. For. Prod. J. 68 (1), 86–95.
Pušavec, F., Kopač, J., 2011. Sustainability assessment: cryogenic machining of Inconel 718. Stroj Vestn-J Mech E 57 (9), 637–647.
Pusavec, F., Kramar, D., Krajnik, P., Kopac, J., 2010. Transitioning to sustainable production–part II: evaluation of sustainable machining technologies. J. Clean. Prod. 18 (12), 1211–1221.
Rentizelas, A., Shpakova, A., Mašek, O., 2018. Designing an optimised supply network for sustainable conversion of waste agricultural plastics into higher value products. J. Clean. Prod. 189, 683–700.
Ross, P.G., 2018. inventor; Amazon Technologies Inc, assignee, Energy production from waste heat using a shape memory drive or heat engine. United States patent US 9,920,750.
Rubinskaya, A.V., Pister, E.I., Ioffe, A.A., Vititnev, A.Y., Karaseva, M.V., 2018. Optimization of recycling strategy for solid industrial waste. In: IOP Conference Series: Materials Science and Engineering. Vol. 450, No. 6. IOP Publishing, p. 062016.
Russ, W., Meyer-Pittroff, R., 2004. Utilizing waste products from the food production and processing industries. Crit. Rev. Food Sci. Nutr. 44 (1), 57–62.
Schill, M., Shaw, D., 2016. Recycling today, sustainability tomorrow: effects of psychological distance on behavioural practice. Eur. Manag. J. 34 (4), 349–362.
Sensoneo, n.d.. https://sensoneo.com/sensoneo-global-waste-index-2019/.
Sheets, J.P., Yang, L., Ge, X., Wang, Z., Li, Y., 2015. Beyond land application: emerging technologies for the treatment and reuse of anaerobically digested agricultural and food waste. Waste Manag. 44, 94–115.
Shi, Q., Yu, K., Kuang, X., Mu, X., Dunn, C.K., Dunn, M.L., Wang, T., Qi, H.J., 2017. Recyclable 3D printing of vitrimer epoxy. Mater. Horiz. 4 (4), 598–607.
Shigaki, N., Akiyama, T., Tsukihashi, F., 2002. Exergy analysis of steel production processes. Mater. Trans. 43 (3), 379–385.
Siddique, R., Khatib, J., Kaur, I., 2008. Use of recycled plastic in concrete: a review. Waste Manag. 28 (10), 1835–1852.
Sigurnjak, I., Brienza, C., Snauwaert, E., De Dobbelaere, A., De Mey, J., Vaneeckhaute, C., Michels, E., Schoumans, O., Adani, F., Meers, E., 2019. Production and performance of bio-based mineral fertilizers from agricultural waste using ammonia (stripping-) scrubbing technology. Waste Manag. 89, 265–274.
Singh, S., Ramakrishna, S., Gupta, M.K., 2017a. Towards zero waste manufacturing: a multidisciplinary review. J. Clean. Prod. 168, 1230–1243.
Singh, N., Hui, D., Singh, R., Ahuja, I.P., Feo, L., Fraternali, F., 2017b. Recycling of plastic solid waste: a state of art review and future applications. Compos. Part B 115, 409–422.
Singhal, D., Tripathy, S., Jena, S.K., 2019. Sustainability through remanufacturing of e-waste: examination of critical factors in the Indian context. Sustain. Prod. Consump. 20, 128–139.
Somakos, L., Chasapidi-Mavroeidi, E.R., Aravossis, K., Tzovaras, S., Glavič, P., Stejskal, B., Szilagyi, A., Doukoulos, T., 2016. Innovative 3D training platform for recycling of waste coming from electric and electronic devices. In: Computer Aided Chemical Engineering. Vol. 38. Elsevier, pp. 2259–2264.
Sreejith, P.S., Ngoi, B.K., 2000. Dry machining: machining of the future. J. Mater. Process. Technol. 101 (1–3), 287–291.
Sun, Z., Cao, H., Xiao, Y., Sietsma, J., Jin, W., Agterhuis, H., Yang, Y., 2017. Toward sustainability for recovery of critical metals from electronic waste: the hydrochemistry processes. ACS Sustain. Chem. Eng. 5 (1), 21–40.

Tischner, U., Hora, M., 2019. Sustainable electronic product design. In: Waste Electrical and Electronic Equipment (WEEE) Handbook. Woodhead Publishing, pp. 443–482.
Tran, T.N., Bayer, I.S., Heredia-Guerrero, J.A., Frugone, M., Lagomarsino, M., Maggio, F., Athanassiou, A., 2017. Cocoa shell waste biofilaments for 3D printing applications. Macromol. Mater. Eng. 302 (11), 1700219.
Veleva, V., Bodkin, G., Todorova, S., 2017. The need for better measurement and employee engagement to advance a circular economy: lessons from Biogen's "zero waste" journey. J. Clean. Prod. 154, 517–529.
Vezzoli, C., Manzini, E., 2008. Design for Environmental Sustainability. Springer, London.
Vourdoubas, J., 2018. Possibilities of using industrial waste heat for heating greenhouses in northern Greece. J. Agric. Sci. 10 (4).
Wei, M.S., Huang, K.H., 2001. Recycling and reuse of industrial wastes in Taiwan. Waste Manag. 21 (1), 93–97.
Welfens, M.J., Nordmann, J., Seibt, A., 2016. Drivers and barriers to return and recycling of mobile phones. Case studies of communication and collection campaigns. J. Clean. Prod. 132, 108–121.
World Bank. n.d. Global waste generation could increase 70% by 2050. https://www.wastedive.com/news/world-bank-global-waste-generation-2050/533031/.
World Wildlife Fund and the Conservation Foundation, 1991. Getting at the Source. Strategies for Reducing Municipal Solid Waste. The Final Report of the Strategies for Source Reduction Steering Committee, Washington, DC, p. x.
Xiong, Y., Zhao, Q., Zhou, Y., 2016. Manufacturer-remanufacturing vs supplier-remanufacturing in a closed-loop supply chain. Int. J. Prod. Econ. 176, 21–28.
Yang, G.C., Chuang, T.N., Huang, C.W., 2017. Achieving zero waste of municipal incinerator fly ash by melting in electric arc furnaces while steelmaking. Waste Manag. 62, 160–168.
Yenipazarli, A., 2016. Managing new and remanufactured products to mitigate environmental damage under emissions regulation. Eur. J. Oper. Res. 249 (1), 117–130.
Yesilata, B., Isıker, Y., Turgut, P., 2009. Thermal insulation enhancement in concretes by adding waste PET and rubber pieces. Constr. Build. Mater. 23 (5), 1878–1882.
Zaman, A.U., 2014. A Comprehensive Review of the Development of Zero Waste Management: Lessons Learned and Guidelines., https://doi.org/10.1016/j.jclepro.2014.12.013.
Zhang, Q., Zhao, X., Lu, H., Ni, T., Li, Y., 2017. Waste energy recovery and energy efficiency improvement in China's iron and steel industry. Appl. Energy 191, 502–520.
Zhang, Z., Han, W., Chen, X., Yang, N., Lu, C., Wang, Y., 2019. The life-cycle environmental impact of recycling of restaurant food waste in Lanzhou. China Appl. Sci. 9 (17), 3608.
Zhou, Z., Dai, G., Hu, C., Zhang, X., 2016. Technology architecture of intelligent remanufacturing. In: 6th International Workshop of Advanced Manufacturing and Automation. Atlantis Press.
Zlamparet, G.I., Ijomah, W., Miao, Y., Awasthi, A.K., Zeng, X., Li, J., 2017. Remanufacturing strategies: a solution for WEEE problem. J. Clean. Prod. 149, 126–136.

Further reading

Debieb, F., Kenai, S., 2008. The use of coarse and fine crushed bricks as aggregate in concrete. Constr. Build. Mater. 22 (5), 886–893.

CHAPTER FOUR

Challenges, issues, and problems with zero-waste tools

K.S. Vignesh[a], Suriyaprakash Rajadesingu[b], and Kantha Deivi Arunachalam[b]
[a]School of Public Health, SRM Institute of Science and Technology, Chennai, India
[b]Center for Environmental Nuclear Research, Directorate of Research, SRM Institute of Science and Technology, Kattankulathur, Tamil Nadu, India

Contents

1 Introduction 69
2 The concept of "zero waste" systems 71
2.1 The holistic model of ZW 73
3 The ZW development 75
3.1 3R model in Bangladesh 75
3.2 ZW development in Spain 77
3.3 ZW development in Australia 77
3.4 ZW development in San Francisco 77
3.5 ZW development in Subaru, US 78
3.6 ZW development in DuPont 78
3.7 ZW development in Coca-Cola 78
3.8 ZW development in India 78
4 The key aspects of developing a ZW strategy 79
4.1 Challenges in ZW management 81
5 Conclusion 87
References 88

1 Introduction

Today's consumer-driven culture creates an immense amount of waste. This vast volume of waste places immense pressure on the municipal authority to handle waste more efficiently. Waste management mechanisms have not gained as much attention in the urban planning process as other sectors such as water or electricity. As a consequence, gaps in waste management can be found in current urban planning. Among all the key challenges, waste management is one of the most important challenges for sustainable

Concepts of Advanced Zero Waste Tools
https://doi.org/10.1016/B978-0-12-822183-9.00004-0

urban design. In high-consumption cities in the industrial world, large amounts of paper wastes, over-packaging, organic waste, chemical waste, industrial waste, food waste, and e-waste are all causing particular socio-economic and environmental problems. "Zero waste" means the design and management of products and processes systematically to prevent and eliminate waste and materials and to conserve and recover all resources from waste streams (Zaman and Lehmann, 2011).

Kuboňová et al. (2013) and BinYousuf and Reza (2013) stated that increasing population, a booming economy, rapid urbanization, and an increase in community living standards have greatly accelerated the generation of solid waste in the world, particularly in developing countries. Solid waste has been one of the major environmental issues. Waste is a sign of the inefficiency of every industrial society and the representation of misdirected capital. The global amount of solid waste is estimated at around 11 billion tons per year (with 2.5 tons of trucks capable of turning 300 circles around the equator) in 2011, and the generation per capita of solid waste is estimated at around 1.74 tons per year worldwide.

On the other hand, along with the large-scale generation of solid waste, an immense amount of natural resources are being consumed every day due to the growing demand for new product production (Kuboňová et al., 2013; BinYousuf and Reza, 2013). Chalmin and Gaillochet (2009) from their findings concluded that globally, 120–130 billion tons of natural resources are consumed annually, producing about 3.4–4 billion tons of urban solid waste. The production of any waste depletes natural resources, consumes energy and water, puts pressure on land, pollutes the atmosphere, and, ultimately, creates increased economic costs for waste management.

Gunningham (1998) reported that this large volume of waste has also generated tremendous pressure on the authority to handle waste more efficiently. Solid waste management is important and significant as the structure of society shifts from a low-density agricultural population to a high-density urban population. Moreover, industrialization has introduced a vast number of goods, the essence of which cannot or can only very gradually decompose or digest. As a result, Guyer (1998) reported that some industrial goods contain substances that, due to low degradability or even toxic characteristics, build up to levels that pose a danger to human future use of natural resources, that is, drinking water, agricultural land, air, and so on.

Ramezani (2015) examined that heavy metals in fly ash from urban solid waste incinerators are found at high concentrations. Fly ash must therefore be handled as a hazardous material. They also conducted fly ash heat treatment in a rotary reactor between 950 and 1050°C and in a muffle oven at temperatures between 500 and 1200°C and stated that more than 90% of the removal was accomplished by simple volatile heavy metals such as cadmium and lead and also by copper, but at higher temperatures in the muffle oven.

In recent times, more attention has been paid to the 3Rs concept due to the depletion of valuable resources and the increase in the amount of pollution in the atmosphere. The outlook for waste management has been changed. It's not seen as a problem but as an opportunity. National 3R waste management policy was developed by the Government of Bangladesh in 2010. It is further supported by the Ministry of Local Government for the implementation of City Corporations and Municipalities Ramezani (2015) stated that the chemical industry is one of the most polluting manufacturing industries. Guyer (1998), from his research findings, concluded that recycling and waste management are the best ways to avoid emissions caused by the chemical sector (Guyer, 1998).

Ramezani (2015) pointed out that the paint industry is one of the most polluting industries in the chemical sector and produces significant volumes of wastewater during cleaning processes (Ramezani, 2015). The researchers Guyer (1998), from their study, specified that painted wastewater is characterized by higher concentrations of styrene-acrylic resins, chemical oxygen demand (COD), biological oxygen demand (BOD), calcium carbonate, titanium dioxide, suspended solids, and colored substances El-Shazly et al. (2010).

Ramezani (2015) carried out the Pollution prevention (P2) assessment by applying the three R's, reducing, reusing and recycling to the chemical industry to reduce the amount of wastewater produced, reusing paint wastewater in the manufacture of cement bricks, recycling of cooling water and improving the efficiency of water usage. The results of their study showed that the annual wastewater flow produced by paint production can be reduced from $1100\,m^3$ to $488.4\,m^3$ (44.4% reduction) when a high-pressure hose is used Mostafa and Peters (2017). The current study aims to identify the challenges, issues, and problems on zero-waste (ZW) tools with respect to socio-economic and environmental health aspects. The present study also highlights the existing challenges faced in the recovery of metals and other constituents that pollute the environment through human activity.

2 The concept of "zero waste" systems

The term "zero waste" was first used by Dr. Paul Palmer in 1973 for the regeneration of chemical resources (Palmer, 2004). In a ZW model, the material flow is continuous, which ensures that the same materials are used again and again until the optimal level of consumption is achieved. No resources are lost or underused in circular structures (Murphy and Pincetl, 2013). As a result, goods are reused, repaired, sold, or recycled throughout the network at the end of their lives. If re-use or repair is not feasible, it may be recycled or extracted from the waste stream and used as inputs to meet the need for the production of natural resources.

ZW is a change from the conventional business paradigm in which waste is considered the standard to integrated systems in which everything is used. It promotes an industrial revolution whereby companies reduce the strain they put on natural resources and learn to do more with what the Earth produces. The ZW principle incorporates the "3R rule" – "Reduce, Reuse, Recycle," and has been used as a foundation for environmental consciousness and a way of fostering ecological sustainability through conscientious action and choices. It is widely agreed that these patterns of behavior and consumer preferences will contribute to reductions in materials and resources that will support the environment.

According to Greyson (2007), the concepts of "Zero Waste" have been implemented in countries like South Africa, New Zealand, China, and India, provinces or states (Nova Scotia, California), as well as a range of companies such as Greyson (2007). The 3R model is a move from the cradle-to-grave approach, which relies on landfills and incineration, to the cost-effective cradle-to-cradle (or circular economy) approach to waste management where waste from one product serves as input for another, negating the need for reliance on landfills and incineration as the only approach for Solid Waste Management (SWM) (Greyson, 2007).

The cradle-to-grave is a linear model for materials that starts with the extraction of resources, passes to the creation of the product, and finishes with a "grave" where the product is disposed of in the landfill. Cradle-to-grave is in a direct contrast to cradle-to-cradle materials or goods that are recycled into goods at the end of their lives, such that there is essentially no waste. Cradle-to-cradle approach focuses on the design of manufacturing processes so that materials flow into closed-loop loops, which ensures that waste is eliminated and waste products can be recycled and reused. Cradle-to-cradle approach goes beyond dealing with issues of waste after it has been made, resolving problems at source and re-defining problems by concentrating on design. The cradle-to-cradle concept is sustainable and respectful of life and future generations (McDonough and Braungart, 2002). The ZW principle has been adopted by policy makers as it promotes sustainable production and use, efficient recycling, and conservation of resources (Zaman, 2014).

2.1 The holistic model of ZW

- In this new holistic model, we have to abandon the aspiration to consume more and more, for example, to purchase more and more products, and start the change toward a low-carbon world. This signifies both making improved, more effective technologies accessible, and also rallying changes in behavior and attitudes. Certainly, 25% of the decrease in discharges needs to come from behavioral change. The new ecological prototype to do business and urban development will be about the integration of systems and activating modernization at all stages. What does the ZW city appear like?
- In the future, we will be renovating existing infrastructure, communities, and construction material; in the meantime, we will develop new ones. We will develop sustainable designs inspired by nature where waste is seen as a resource and organic waste is used as a fertilizer and where new building materials are created from recycled waste.
- In the future, energy will be generated from potential renewable energy sources like wind, geothermal, solar, and biomass resources. We will change the way we generate energy and see more and more decentralized systems on roofs and facades, where cities become power stations in themselves, and where all citizens can become energy producers (instead of just being consumers).

In the future, the ZW cities will be developed that produces fewer waste, by collecting waste, complete recycling and recovery of resource, and guaranteeing sustainable resource usage. To attain the objectives of ZW city, five interlinked vital principles that have to be applied concurrently for converting a city into a ZW city are proposed. The five key principles proposed for converting a city into a ZW city are described below

1. change in behavior and sustainable utilization,
2. extended producer and consumer accountability,
3. complete recycling of municipal solid waste,
4. legislated zero incineration and landfill, and
5. complete recovery of resources from waste.

The above-mentioned five principles are the vital converters to transform cities into ZW cities. Apart from that, all the above-said five principles must be applied concurrently to attain effective results in the transformation process (Allen, 2012). However, depending on the ensemble application of each principle, a long-term ZW city vision would be required because the concept of a ZW city vision is not only very hard to achieve but also requires long-term initiatives to achieve that. Fig. 1 shows the complete principles to transform a city into a ZW city. The principles of the ZW city are established based on waste hierarchy, that is, evade, reduction, and retrieval. Change in behavior and sustainable utilization practice will evade the unwanted generation of waste from product manufacture and use stages. Extended producer and consumer accountability will confirm the sustainable option of resource utilization and possession of individual waste generation and management. An amplified sense of accountability will also evade the generation of waste. Resource and product stewardship would reduce the impacts on the environment in the long run and assure the welfare of the future cohorts by defending resources through a behavior change from over-usage to

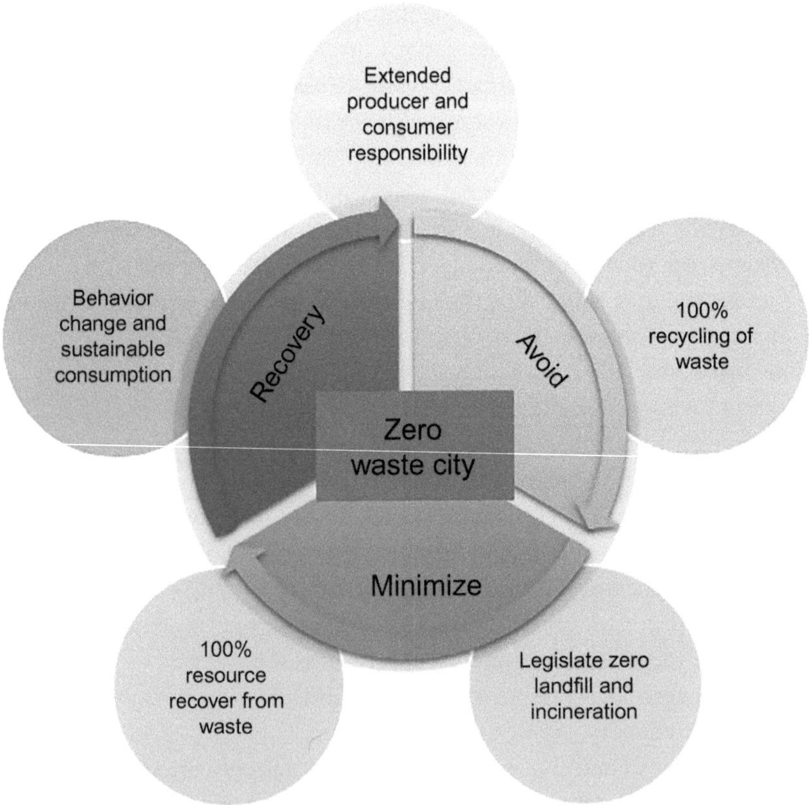

Fig. 1 A holistic zero waste city model.

practical and sustainable usage (Hussain and Kharisov, 2017). By reaching complete waste recycling and legislation for zero incineration and landfill, complete resource recovery would be likely to happen in the ZW city and therefore ensuring the least depletion of limited natural resources.

In the strategy of ZW city, existing cities have to be re-constructed to attain more sustainable and adaptable (Qingbin Song Jinhui, 2014). From the usage of fossil fuels to technologies with low-carbon emission, we will transform and redesign the method we plan, build, function, recycle constructions, neighborhoods, and finally cities. This necessitates us to ponder about many things in a different way than we had in the past, for example about our emissions exhausting production bases, our extravagant supply chains, and most importantly our old-fashioned material-ineffective building methods. In this evolution, some cities and industrial sectors will be leaping ahead, whereas others may be at risk of being left behindhand. The waste management sector has some of the utmost prospects to reinvent itself.

3 The ZW development

ZW hierarchy defines the progression of policies and approaches to achieve ZW from the highest and greatest to the lowest usage of materials. It is designed to be accessible to all audiences, from policy makers to industry and individuals. This seeks to provide more scope to globally accepted 3Rs (Reduce, Reuse, Recycle); to promote policy, action, and investment at the top of the hierarchy and to guide those who wish to build programs or goods that bring us closer to ZW. Globally, the ZW concept has been implemented through various strategies.

3.1 3R model in Bangladesh

In Bangladesh, the process of recovery and recycling occurs in three phases.

Phase 1: Waste generators detach waste with higher market value, such as paper, bottles, and plastic containers, and sell them to street hawkers.

Phase 2: Scavengers are rummage through waste in the vicinity of bins for the accumulation of recyclable materials of low market value, such as broken glass, bottles, and polythene discarded by households.

Phase 3: Collection of recyclable materials by waste pickers from waste vehicles directly after unloading at landfill sites. From an economic and social point of view, it saves on the use of energy, reduces the cost of waste disposal, and leads to environmental protection.

The availability of funds to help segregation and recycling of waste is a challenging problem. Municipalities are only able to manage basic waste collection and disposal systems, which are heavily funded by the state. Pilot and demonstration projects will play a significant role in complementing national 3R strategies and policies by educating the general public, the private sector, and other key stakeholders on the benefits and impacts of the 3Rs.

3.1.1 Pilot study in Bangladesh

Ministry of Environment and Forest (MOEF) using the Climate Change Trust Fund initiated a demonstration project of 3R (First phase) in 4 communities in Dhaka and 2 communities in Chittagong. The main purpose of the project is to create awareness on source segregation and recycling of waste and reduction of emission of Green House Gases from waste.

3.1.2 Role of stakeholders in 3R

Individuals, families, community/neighborhood (clubs), policy makers (principal/official), business groups (SMEs, manufacturers), informal sectors (waste pickers, secondary buyers), the private sector, CBOs/NGOs, research and academic organizations, and so on are potential partners for the development and implementation of the 3R Action Plan. Based on the levels, each stakeholder plays a unique and important role in ZW development, such as developing policies and guidelines, providing financial and technical resources to promote 3R for related projects, by improving the health and safety of the workers, by implementing awareness programs among the community and involving public into the waste management activities, conducting academic and scientific research programs, to improve and implement various technologies in 3R model.

3.1.3 3R Partnership building between stakeholders

To put 3R activities into action in society, there should be an involvement by all participants and involvement that could be ensured through the establishment of three 3R organization, such as

- 3R promoters (official stakeholder network),
- 3R supporters (community-based volunteer network), and
- 3R volunteers (young generation volunteer network).
- In addition, 3R units should be set up in the city companies to assist and enforce 3R activities.

- Mass media (Newspaper, Electronic Media) will engage on a daily basis in favor of 3R events.

3.2 ZW development in Spain

The government has implemented the door-to-door collection program in the city of Hernani, Spain. Hernani and two other small towns in the area have set up an ambitious system of door-to-door recycling of waste, including organic waste, which has been enthusiastically welcomed by citizens. The volume of waste to the landfill was decreased by 80% and the rate of waste diversion was 79%. Public investment per capita in solid waste management amounted to US $115 per year With a new political leader opposed to incineration, door-to-door collection is expected to grow across the country (Hussain and Mishra, 2019).

3.3 ZW development in Australia

Adelaide is the capital city of South Australia and includes 19 municipal areas. A total of 1,089,728 inhabitants live in an urban area of 841.5 km^2 (Task Force on Waste to Energy, 2014). Adelaide City Council (ACC) is responsible for the disposal of waste in Adelaide. ZWSA is a Southern Australian state government agency founded by the ZWSA Act (2004). ZWSA encourages people to improve their recycling and waste management activities at home, at work, and in the industry. Adelaide has a high proportion of waste collection systems relative to other Australian capital cities. Container deposit legislation was introduced in 1977; thus, recycling of various shipping containers has been practiced for more than three decades. ZWSA is working toward a ZW environment in South Australia.

3.4 ZW development in San Francisco

Zaman and Lehmann (2011), from their research finding, stated that, San Francisco has developed itself as a global leader in waste management. The city has achieved 77% waste diversion, the highest in the United States, with a three-pronged approach:

- Introducing strict waste reduction regulations,
- Collaborating with a similar waste management company to develop new technologies, and
- Working to build a recycling and composting community through rewards and outreach.

3.5 ZW development in Subaru, US

- Subaru USA has done a fantastic job at their Lafayette site in Indiana, and Subaru estimates that 95% of their waste is being recycled or recycled through the following activities (Qingbin Song Jinhui, 2014);
- Copper-loaded slag left over from welding is gathered and sent to Spain for recycling;
- Styrofoam types enclosing delicate engine parts are returned to Japan for the next round of deliveries; and
- Also, tiny protective plastic caps are stored in bins to be melted down to create something special.

3.6 ZW development in DuPont

In 2012, the DuPont Construction Technologies carried out "Zero Landfill" and was able to find a way to recycle much of the products from the Corian site (Hansen et al., 2002):

- Scrap sheet and trim—ground back to 1st level or ground to rock,
- Carrier film—turned into glue,
- Metal—melted and recast,
- Banding—melted and recast,
- Pallets—repaired and reused,
- Scrap wood—grounded to animal bedding, and
- Paper and cardboard—reused as same.

3.7 ZW development in Coca-Cola

Coca Cola claims that they have done a lot of beneficial research in ZW development, such as:

- Trimming the weight of their 20-oz Polyethylene terephthalate (PET) plastic bottle by more than 25%,
- From 12-oz aluminum tin about 30% of its weight is reduced, and
- Lightening their 8-oz glass bottle by more than 50%.

In 2011/2012, Coca Cola reported that they saved $180 million from reducing their packaging.

3.8 ZW development in India

India's adoption of ZW as a strategy for SWM has primarily been driven at the local level by local governments such as Pune and Mysore, at the state-government level by states such as Gujarat, or NGOs such as Chintan, ExNoRa, and Thanal.

In 2011, urban India produced 62 million tons of solid waste per year (Task Force on Waste to Energy, 2014) which is expected to increase by 5–260 million tons per year by 2047 (TERI). Under the Swachh Bharat Mission (SBM), 80% of the urban population (with a rise of 2% year-on-year) must be covered by the Solid Waste Management (SWM) services.

It is therefore imperative that environmentally sustainable, socially, and techno-economically viable SWM activities are implemented to ensure sustainability in the long term. The Ministry of Environment and Forestry (MoEF) drafted the SWM rules in 2000. The 2015 draft amendments recommend that every attempt should be made to achieve the desired ZW objective; landfills should be permitted only for non-usable, non-recyclable, non-biodegradable, non-combustible, and non-reactive waste. The ZW strategy is based on 3Rs in India.

ZW as a term had its roots in the effectiveness concepts developed in Japan, focusing on the ideas of total quality management (TQM) and zero defects that enabled producers like Toshiba to achieve results as low as one defect per million units produced. It has been extended to approaches to deal with industrial waste, wastewater, and urban solid waste. ZW experiments are common in Northern Europe, the US, and other countries such as New Zealand, and India's experiments are typically carried out on a pilot scale across the world, such as Kovalam, ExNoRa Vellore, Chintan, Pune, Mysore, Ahmedabad, and others. These are locally guided projects, either by citizens' groups/NGOs and/or local governments (ULBs or district authorities).

As the second largest city in Maharashtra, Pune generates 1600 tons of garbage every day. The rising quantity and difficulty of waste handling pose a serious challenge for India to Swach by 2019. Rapid urbanization, growing population, and spatial constraints have led to the quest for creative and sustainable solutions to make Pune a garbage-free city. "Zero Garbage Pune" is an ambitious initiative for sustainable waste management in conjunction with the Swachh Bharat Campaign. In Pune, the waste is collected, stored, and separated for recycling and re-use. The remainder shall be transported for further processing and scientific disposal.

4 The key aspects of developing a ZW strategy

The 3R principles (reduction, re-use, and recycling) are among the top three in the waste hierarchy, and they are considered as the founding principles of the sustainable waste management system (Hansen et al.,

2002). In the European Union Waste Framework Directive 2008, the "3R" principles have been extended to five steps of the waste hierarchy: prevention (avoidance), re-use, recycling, recovery (including energy recovery), and disposal (European Commission Preparing a Waste Prevention Programme, 2012). A variety of strategies have been established in different studies, such as eco-design, conscientious shopping behavior, and so on, in relation to the prevention and reduction of waste (Braungart et al., 2007; Schmidt, 2012).

Waste reduction is one of the most critical challenges in ZW and needs mutual social consciousness and expertise on waste and creative development and business models (Cox et al., 2010). Awareness and transformative knowledge are also thought to inspire behavioral change in relation to environmental lifestyle choices (Jackson, 2005). Responsible and productive customer conduct is another critical concern in the avoidance of waste.

Collaborative consumption increases the efficiency of resource use and improves social collaboration Rogers and Botsman (2010). Collaborative ownership or collaborative consumption models encourage service-based enterprise and waste prevention. The re-circulation (circulation of materials in the supply chain for repeated use) of post-consumer goods by re-use and re-selling is therefore important and boosts the circular economy and increases social capital. Waste management and treatment techniques have been used to address waste issues for more than a century (UNEP/GRID-Arendal, 2006).

ZW takes the view that technology alone can't address waste issues sustainably, as it needs community participation, service infrastructure, regulatory policy, and environmentally sustainable treatment technology. Several studies have established that efficient collection systems, decentralized waste recycling centers, social innovations, such as recycling, composting, regulatory policies such as pay-as-you-go (PAYT) and environmentally sustainable advanced waste management innovations, are key issues in waste management and treatment (Dahlén and Lagerkvist, 2010; Seyfang, 2005; Serpe et al., 2015). The basic differences between conventional waste management and ZW management are that it limits the use of waste-to-energy waste (WTE) that burns waste to produce energy (heat and electricity) and landfills in an "ideal" ZW environment. The ZW strategy has five successful key aspects

1. segregating waste at the source,
2. creating awareness among public,
3. the process of decentralization of waste,

4. community composting of waste, and
5. incorporation of informal economy in the ZW management.

Dr.Chaturvedi in the NITI Aayog—Centre for Policy Research (CPR) Open Seminar Series proposed two paths, that is, techno-nirvana and green transformations to achieve ZW in India. Techno-nirvana focuses on extracting the full value from waste by large-scale, capital-intensive technology. This includes PPP and technology-based approaches. Challenges on this route require heavy investment and higher emission control costs.

4.1 Challenges in ZW management

Both global economic growth and consumption rate have increased significantly all around the globe Rogers and Botsman (2010). Waste generation trends indicate that waste volume reduction is one of the key challenges for all cities (Ackerman, 2007; Sinha and Amin, 1995). High-consuming cities, such as San Francisco, Copenhagen, and Stockholm, have been implemented different methods and policies to collect 100% of waste from the source of generation and to manage it properly. These cities are quite successful in the diversion of waste from landfill (Leach et al., 1997; Huang et al., 2006; Evans, 2011). However, all these cities are facing problems in the context of long-term sustainable resource recovery. Sweden and Denmark were incinerated recovery sustainable energy and heat by the municipal waste around 50% in 2009. Conversely, incineration exhausts valuable resources like plastic, paper, cloths, and different materials. It can be reused and recycled from various techniques.

In developing countries like India and China, the volume of waste materials is increasing over time parallel due to population growth in the city and rural areas. Consumption of resources has been increasing in China significantly in the last few decades, which indicates potential increasing rates of waste generation in low-consuming cities. Therefore, taking consideration of low consuming city contexts, where intake level has been raised and landfill is the main waste treatment technology, developing waste management in cities is also very challenging to achieve sustainably. The global waste management and finite resource scenario will be more difficult to manage when low consuming countries reach the same consumption rates as the high consuming countries.

The study also acknowledged that cities are very dynamic and combine different complex spheres. Moreover, cities' population lifestyle is different from other rural population due to geographical and environmental factors.

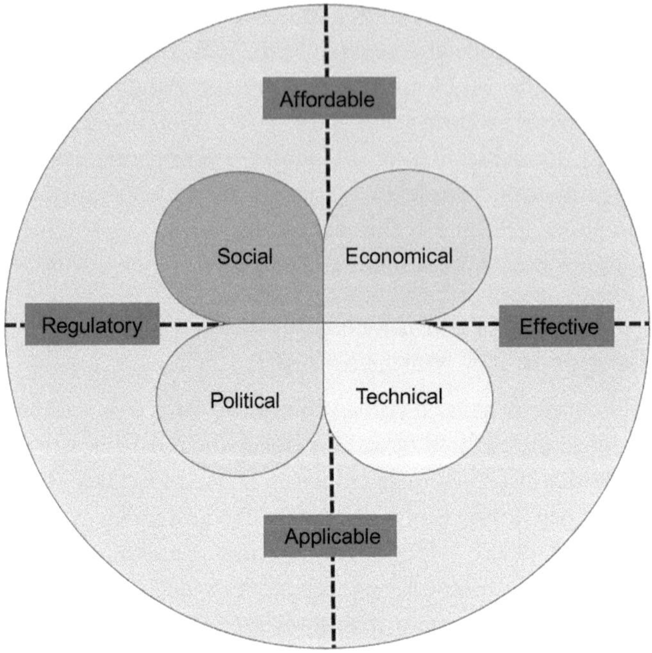

Fig. 2 Spheres in a sustainable zero waste city.

Subsequently, the problem of city and rural is not easy to understand the dynamic nature of the factors involved in sustainable way in the city development without holistic research approaches. Fig. 2 shows the complexity in designing ZW cities, where the environmental sphere works as a rim for all other spheres such as social, economic, political, and technological, and all those spheres are dynamic.

This chapter recognized five core aspects that are most important in converting cities into ZW and sustainable cities. The tools, methods, technology, or strategies developed for recycling or dealing waste in ZW cities should be reasonable in the socio-economic context, regulatory or manageable in the socio-political context, applicable in the policy and technological context, effective or efficient in the context of economy and technology, and finally, all these features should be straightly related to environmental sustainability.

4.1.1 Challenges in the recovery of industrial waste

Industrial waste is created by industrial activities, such as factories, mills, and mines. Even, there is still a large portion of solid waste. Total generation in

Table 1 Generation of industrial waste in 2011.

Items	Europe	America	Africa and the Middle East	Asia Pacific	Global
Generated (million tons)	1933.19	914.7	921.24	5357.46	9176.68
Collected (million tons)	1531.71	765.8	270.91	3346.02	5914.5
Key countries	Germany, UK, France, Russia, and Bulgaria	US, Brazil, Canada, Chile, and Columbia	South Africa, Saudi Arabia, UAE, Egypt, and Tunisia	China, Japan, India, South Korea, and Australia	–

Source: The authors..

2011 was approximately 9.2 billion tons (including construction waste) (refer Table 1), and per capita industrial waste in the world is about 1.74 tons per year. As shown in Table 1, more than 50% has been produced in Asia and the Pacific, particularly in China (3.2 billion tons in 2011; Zeng et al., 2010). Industrial waste in developing countries (such as China) has a fast-growing trend in the near future, while industrial waste generation appears to be stable or slow to decline (such as in Spain) in developed countries (Frost and Sullivan, 2012). In comparison, relative to developed countries (more than 90% collection rate), many developing countries remain at a lower collection and recycling rate (China 67% in 2010; India less than 50% in 2010). Developing countries are also facing more threats from toxic waste.

4.1.2 Environmental concern

As far as industrial waste (especially hazardous waste) is concerned, illegal dumping and trans-boundary movements often attract more attention due to their potential risks to the environment and human health. Illegal dumping is a chronic problem as it endangers human health and the environment, imposes substantial costs on communities, and has a detrimental impact on public wellbeing (Penelope et al., 2010). Illegal dumping also leads to more illegal dumping. Owing to the large generation of industrial waste, not all waste can be collected and processed, thus enormous industrial waste has been illegally dumped, especially in developing countries (Penelope et al., 2010).

Industrial waste producers are constantly confronted with the question of disposing of their waste and must choose from a variety of different disposal and treatment options. They are slowly opting to distribute waste to other countries (most of them are developing countries) (Chartsbin, 2013). It is very difficult to assess how much hazardous waste is transported every year. The trans-boundary movement refers not only to the export of waste but also to the transfer of environmental pollution from developed nations to developing nations.

Many significant environmental disasters have occurred as a result of illegal mining and trans-boundary movements. In 2011, the authors Maantay and McLafferty reported that a Chemical Mining firm illegally dumped 5222.38 tons of chromium slag in Yunnan Province. In 2006, a ship registered in Panama, Probo Koala, chartered by the Dutch oil and commodity corporation Trafigura Beheer BV, unloaded radioactive waste at the Ivorian port of Abidjan. The various researchers explored the reasons why illegal dumping and trans-boundary movements have occurred, primarily related to possible economic benefits, for example, the cost to a European corporation of proper disposal of hazardous waste would cost about $1000 per ton, if the materials were illegally dumped, the cost would be around $2.50 per ton; the cost of legally incinerating waste in the Netherlands is four times higher.

4.1.3 Challenges with e-waste recovery

Electronics are complex devices consisting of a wide range of material components. These unwanted electronic goods consist of plastics, metals, glass, and other materials and chemicals, some of which are harmful. If they are wrongly treated, hazardous compounds, such as polychlorinated dibenzo-p-dioxins and dibenzofurans (dioxins; PCDD/Fs), flame retardants, and heavy metals (e.g. lead, mercury, arsenic, cadmium, selenium, and hexavalent chromium), can be emitted and may cause significant environmental contamination.

The majority of electronic gadgets are sold in developed countries, such as the United States, Japan, Australia, and Europe, and it is estimated that between 50% and 80% of the e-waste collected for recycling in the United States is exported to less developed countries (Kojima et al., 2013). It has also been found that the United Kingdom is the leading European exporting country, followed by France and Germany (Basel Convention). Some regions have become hubs for informal e-waste recycling, the most prominent in Asia and Africa.

Informal electronic waste recycling has also been recorded in Qingyuan, Guiyu, Fengjiang, Zhejiang Region, Guangdong Region, China, Hong Kong, Philippines, Manila, Bangalore, Chennai, India, South Korea, Seoul, New Territories, and Hai Phong District, Dong Mai and Bui Dau, Vietnam. Staff and local residents are exposed to harmful chemicals through inhalation, dermal absorption, ingestion of dust, and oral intake. Inhalation and dust ingestion present several possible occupational hazards, including silicosis. Overall, human health hazards from e-waste include trouble breathing, coughing, respiratory distress, diarrhea, vomiting, neuropsychiatric issues, tremor, epilepsy, coma, and even death (Kojima et al., 2013).

4.1.4 Challenges in recycling of metals

Vandevivere et al. (2001) used strong metal chelant (S, S; stereoisomer of ethylenediamine disuccin) to extract Pb, Zn, Cu and Cd from actual polluted media, while Andreottola et al. (2010) researched the use of electrochemical oxidation, chemical oxidation and electrokinetics under different conditions for the remediation of dredging materials from Venice Lagoon. Yoo et al. (2013) also focused on dredged marine bottom sediments and the feasibility of using chemical complex washers to remove metals. Studies on the remediation of metals from bay sediments and key parameters that play an important role in process efficiency were performed in the laboratory by Zhang et al. (2010) in which chemical reagents such as sodium dodecylsulphate (SDS), oxalic acid ($H_2C_2O_4$), ethylenediaminetetraacetic acid, disodium salt (EDTA-2Na), acetic acid, ammonium acetate ($N_2C_2O_4$) were used.

Several experiments have been performed on harbor sediments, and The researchers from multi-disciplinary domain (2009) have investigated the use of electrodyalitic processes as an extraction technique for the treatment of Cu, Zn, Pb, and Cd firmly bound to anoxic sediments on a laboratory scale. Several studies have relied on the possibilities for extraction/removal of metals from ash and slag generated in municipal solid waste incineration plants through various methods, including thermal and hydro-metallurgy (Kuboňová et al., 2013), dry discharge of bottom ash, accompanied by a series of magnetic and Eddy currents, fly ash acid leaching, multi-stage dust collection methodology, and a novel hybrid method using bio-electrochemical systems (BES) followed by electrolysis reactions (ER) (Tao et al., 2014). The application of high-intensity wet magnetic separation to extract As, Cu, Pb, and Zn from sandy loam soil from the mining site was investigated by Sierra et al. (2014). Cu extraction from artificially polluted sediments by the use of an electro-kinetic method powered by a galvanic cell has also been published.

Several authors have researched various methods for removing metals from solid-phase materials. However, there is no literature and thus a lack of information when considering the fine-grained section of the excavated landfill, the bottom sediments of heavily polluted ports, and the potential for metal recovery. The ESEG group at LNU has begun work on the availability and composition of metals from both Oskarshamn Harbor (Fathollahzadeh et al., 2014) and Swedish and Estonianquarried landfills. Initial findings have shown that nearly 50% and 40% of excavated waste volumes are below 40mm and 10mm, respectively. In addition, preliminary studies performed by LNU have shown that metals such as zinc, copper can be plentiful, suggesting further studies on how to isolate and obtain the correct fractions of each metal.

4.1.5 Challenges in the recovery of chemicals

The feasibility of extracting metals from multi-contaminated fine-grained fragments in landfills is dependent on material structure, particle size distribution, and also metal concentrations. For example, hydrometallurgy may be used to extract metals from low-grade materials in which dispersed metals are selectively dissolved into solution. The solubility of the metals and the correct selection of the eluent (which can be recycled) play a significant role in the process. The key emphasis must now be on comparative studies to determine the feasibility of using specific desorbing agents for selective metal recovery, for example:

a. the Chemical leaching process (cyanide leaching, thiourea leaching, halide leaching, thiosulphate leaching) and
b. the chemical leaching including ligands (DTPA, EDTA, Oxalate, and NTA,), and
c. chemical leaching by acid treatment such as hydrochloric acid, sulfuric acid, nitric acid, aquaregia, and sodium hypochlorite.

Apart from chemical leaching processes, it is important to consider some experimental variables if the main objective is to gain a better understanding of how successful a particular metal recovery process is. The parameters that can be used in the experimental design include aeration, agitation speed (rotation), initial pH, and liquid-to-solid ration, provided that these variables would also imply economic viability, which will be of great importance in the experimental methods (Hussain and Keçili, 2019).

4.1.6 Challenges in electricity based recovery

Various studies have been conducted for evaluating the probability of employing electrolysis processes and electro chemicals in the process of recovering the metals. Various advantages, such as simple experiment protocols, compact instrument, and Linnaeus ECO-TECH´14, Kalmar, Swede high removal efficiency (Tao et al., 2014) makes this a better option for recovering metals present in wastewater. However, researchers uphold that although the electrolysis procedures have been studied for the metal recovery process, the main focus was on the recovery mechanism instead of using in the complex media like fine-grained fractions from landfills, fly ashes, and bottom sediments that are usually contaminated with several metals.

Therefore, the present challenges are on conducting the experiments in the lab and pilot scale through electricity-based methodologies and to optimize it based on the obtained results to gain information about key parameters like contact time, liquid–solid ratio, cell compartments, current densities, exchange membranes, and voltages specific applied to each metal. Also, it is imperative to highpoint that researches must be carried on in materials from various origins and properties (water content, impurity concentrations, grain size distribution, and organic content) to validate the results (Hussain, 2020).

5 Conclusion

As a result of rapid economic growth and globalization, a large amount of solid waste have been created and attracted global attention due to potential environmental impacts and resource waste, such as illegal dumping and transboundary movement of industrial waste, informal recycling of e-waste, loss of food and greenhouse gas emissions, resource use of excess packaging. Today, we are faced with more serious solid waste problems than the past, and how to tackle solid waste problems is a big problem for the world as a whole.

The ZW principle is an important way to address the issues of solid waste. ZW is intended to facilitate the redesign of resource life cycles so that all items are reused. To better understand ZW, the standard ZW activities are discussed in this report. At present, many significant initiatives have been made in towns, businesses, and individuals, which provide us with many strong ideas for realizing ZW in the future.

While there are many zero practices and ZW solutions in the world today, ZW management includes socio-economic, political, environmental, and technical issues and has many stakeholders. ZW is a very complicated program, and there are still tons of works to be done in the future. To eliminate solid waste to ZW, all people will need to work to reduce, reuse, and recycle waste from our land. ZW is also a beneficial objective. However, while the path to ZW may have already been forged, there is still a long road ahead of us.

References

Ackerman, F., 2007. Materials flows for a sustainable city. Int. Rev. Environ. Strateg. 5, 499–510.

Allen, C., 2012. Hernani, "Spain: Door-to-Door Collection as a Strategy to Reduce Waste Disposal". http://no-burn.org/downloads/ZWpercent20Hernani.pdf.

Andreottola, G., Bonomo, L., De Gioannis, G., Ferrarese, E., Muntoni, A., Polettini, A., Pomi, R., Saponaro, S., 2010. Lab-scale feasibility tests for sediment treatment using different physico-chemical techniques. J. Soils Sediments 10 (1), 142–150.

BinYousuf, T., Reza, A., 2013. 3R (Reduce, reuse and recycle) action plan for the city corporations in Bangladesh: paradigm shift of waste management to resource management. In: Proceedings of the Waste Safe 2013 – 3rd International Conference on Solid Waste Management in the Developing Countries, 10–12 February 2013, Khulna, Bangladesh, Paper ID: 122.

Braungart, M., McDonough, W., Bollinger, A., 2007. Cradle-to-cradle design: creating healthy emissions— a strategy for eco-effective product and system design. J. Clean. Prod. 15, 1337–1348.

Chalmin, P., Gaillochet, C., 2009. From Waste to Resource: An Abstract of World Waste Survey 2009. Available from: https://www.veoliaenvironmentalservices.com/veolia/ressources/files/1/927,753,Abstract_2009_GB-1.pdf.

Chartsbin, 2013. Global Illegal Waste Dumping by Country. http://chartsbin.com/view/576.

Cox, J., Giorgi, S., Sharp, V., Strange, K., Wilson, D.C., Blakey, N., 2010. Household waste prevention—a review of evidence. Waste Manag. Res. 2010 (28), 193–219.

Dahlén, L., Lagerkvist, A., 2010. Pay as you throw: strengths and weaknesses of weight-based billing in household waste collection systems in Sweden. Waste Manag. 2010 (30), 23–31.

El-Shazly, M.A., Ezzat, A.H., Kamel, K.M., 2010. Appropriate technology for industrial wastewater treatment of paint industry. Am.-Eurasian J. Agric. Environ. Sci. 8 (5), 597–601.

European Commission Preparing a Waste Prevention Programme, 2012. Guidance Document. European Commission, Paris, France.

Evans, D., 2011. Consuming conventions: sustainable consumption, ecological citizenship and the worlds of worth. J. Rural. Stud. 27, 109–115.

Fathollahzadeh, H., Kaczala, F., Bhatnagar, A., Hogland, W., 2014. Speciation of metals in contaminated sediments from Oskarshamn Harbor, Oskarshamn, Sweden. Environ. Sci. Pollut. Res. 21, 2455–2464.

Frost, Sullivan, 2012. The Global industrial waste recycling & services markets. In: TEKES Growth Workshop in Helsinki on 2nd October 2012. https://tapahtumat.tekes.fi/uploads/c8ffe124/TekesGG_Workshop_021012_global_industrial_waste_presentation-9175.pdf.

Greyson, J., 2007. An economic instrument for zero waste, economic growth and sustainability. J. Clean. Prod. 15 (13), 1382–1390. https://doi.org/10.1016/j.jclepro.2006.07.019.

Gunningham, N., 1998. Environmental management systems and community participation: rethinking chemical industry regulation. J. Environ. Law Policy 16 (2), 319–439.
Guyer, H.H., 1998. Industrial Processes and Waste Stream Management. John Wiley & Sons, New York, NY.
Hansen, W., Christopher, M., Verbuecheln, M., 2002. EU Waste Policy and Challenges for Regional and Local Authorities. Ecological Institute for International and European Environmental Policy, Berlin, Germany.
Huang, Q., Wang, Q., Wang, Q., Dong, L., Xi, B., Zhou, B., 2006. The current situation of solid waste management in China. J. Mater. Cycles Waste Manag. 8, 63–69.
Hussain, C.M., 2020. The Handbook of Environmental Remediation: Classic and Modern Techniques. Royal Society of Chemistry.
Hussain, C.M., Keçili, R., 2019. Modern Environmental Analysis Techniques for Pollutants, first ed. Elsevier.
Hussain, C.M., Kharisov, B., 2017. Advanced Environmental Analysis-Applications of Nanomaterials. The Royal Society of Chemistry.
Hussain, C.M., Mishra, A.K., 2019. Nanotechnology in Environmental Science, 2 Volumes. John Wiley & Sons.
Jackson, T., 2005. In: Jackson, T. (Ed.), Motivating Sustainable Consumption: A Review of Evidence on Consumer Behaviour and Behavioural Change. Sustainable Development Research Network, Surrey, UK.
Kojima, M., Yoshida, A., Sasaki, S., Chung, S., 2013. Lessons learned from illegal transboundary movement of hazardous waste in Asia. In: International Trade in Recyclable and Hazardous Waste in Asia, September 30, 2013. Edward Elgar Publishing.
Kuboňová, L., Langova, S., Nowak, B., Winter, F., 2013. Thermal and hydrometallurgical recovery methods of heavy metals from municipal solid waste fly ash. Waste Manag. 33, 2322–2327.
Leach, M.A., Bauen, A., Lucas, N.J.D., 1997. A systems approach to materials flow in sustainable cities: a case study of paper. J. Environ. Plan. Manag. 40, 705–723.
McDonough, W., Braungart, M., 2002. Cradle to Cradle: Remaking the Way We Make Things, first ed. North Point Press, New York, NY.
Mostafa, M.K., Peters, R.W., 2017. Applying the three R's: reduce, reuse, and recycle in the chemical industry. J. Air Waste Manage. Assoc. 67 (3), 322–329. https://doi.org/10.1080/10962247.2016.1234421.
Murphy, S., Pincetl, S., 2013. Zero waste in Los Angeles: is the emperor wearing any clothes? Resour. Conserv. Recycl. 81, 40–51. https://doi.org/10.1016/j.resconrec.2013.09.012.
Palmer, P., 2004. Getting to Zero Waste. Purple Sky Press, Sebastopol.
Penelope, C., Tara, M., Kim, W., Alicia, P., 2010. Illegal Dumping and Crime Prevention: A Case Study of Ash Road. Liverpool Council.
Qingbin Song Jinhui, L., 2014. Minimizing the increasing solid waste through zero waste strategy. J. Clean. Prod. https://doi.org/10.1016/j.jclepro.2014.08.027.
Ramezani, L., 2015. Waste minimization in solvent-based paint industries. Nova J. Eng. Appl. Sci. 4 (1), 1–5. https://doi.org/10.20286/jeas.v4i1.32.
Rogers, R., Botsman, R., 2010. What's Mine Is Yours: The Rise of Collaborative Consumption. New York, NY, HarperBusiness.
Schmidt, M., 2012. Retail shopping lists: reassessment and new insights. J. Retail. Consum. Serv. 19 (1), 36–44.
Serpe, A., Rigoldi, A., Marras, C., Artizzu, F., Mercuri, M.L., Deplano, P., 2015. Chameleon behaviour of iodine in recovering noble-metals from WEEE: towards sustainability and "zero" waste. Green Chem. 2015 (17), 2208–2216.
Seyfang, G., 2005. Shopping for sustainability: can sustainable consumption promote ecological citizenship? Environ. Politics 2005 (14), 290–306.

Sierra, C., Martínez-Blanco, D., Blanco, J.A., Gallego, J.R., 2014. Optimization of magnetic separation: a case study for soil washing at a heavy metals polluted site. Chemosphere 107, 290–296.
Sinha, M.M.H.A., Amin, A.T.M.N., 1995. Dhaka's waste recycling economy: focus on informal sector labour groups and industrial districts. Reg. Dev. Dialogue 16, 173–195.
Tao, H.-C., Lei, T., Shi, G., Sun, X.-N., Wei, X.-Y., Zhang, L.-J., Wu, W.-M., 2014. Removal of heavy metals from fly ash leachate using combined bioelectrochemical systems and electrolysis. J. Hazard. Mater. 264, 1–7.
Task Force on Waste to Energy, 2014. Report of the Task Force on Waste to Energy: In the Context of Integrated MSW Management.
UNEP/GRID-Arendal, 2006. A History of Waste Management. http://maps.grida.no.go/graphic/a-history-of-waste-management. (accessed on 10 May 2016).
Vandevivere, P., Hammes, F., Verstraete, W., Feijtel, T., Schowanek, D., 2001. Metal decontamination of soil, sediment, and sewage sludge by means of transition metal chelant [S, S]-EDDS. J. Environ. Eng. 127 (9), 802–811.
Yoo, J.-C., Lee, J.-D., Yang, J.-S., Baek, K., 2013. Extraction characteristics of heavy metals from marine sediments. Chem. Eng. J. 228, 688–699.
Zaman, A.U., 2014. A comprehensive review of the development of zero waste management: lessons learned and guidelines. J. Clean. Prod. 91, 12–25. https://doi.org/10.1016/j.jclepro.2014.12.013.
Zaman, A.U., Lehmann, S., 2011. Challenges and Opportunities in Transforming a City into a 'Zero Waste City'. pp. 73–93.
Zeng, X., Sun, Q., Huo, B., Wan, H., Jing, C., 2010. Integrated solid waste management under global warming. Open Waste Manag. J. 2010 (3), 13–17.
Zhang, D., Tan, S., Gersberg, R., 2010. Municipal solid waste management in China: status, problems and challenges. Environ. Manag. 91, 1623–1633.

CHAPTER FIVE

Application of poultry industry waste in producing value-added products—A review

Sathish Kumar Karuppannan, Mohammed Junaid Hussain Dowlath, G.I. Darul Raiyaan, Suriyaprakash Rajadesingu, and Kantha Deivi Arunachalam
Center for Environmental Nuclear Research, Directorate of Research, SRM Institute of Science and Technology, Kattankulathur, Tamil Nadu, India

Contents

1 Introduction 91
2 Poultry processing steps and wastes generated 93
3 Zero-waste approach in handling poultry waste 95
 3.1 Feather waste 95
 3.2 Hatchery waste 103
 3.3 Poultry litter and manure 105
 3.4 Egg shell waste 110
 3.5 Spent hens 113
 3.6 Offal waste 113
4 Conclusion 114
Important websites 114
Reference 115

1 Introduction

Poultry (Fig. 1) is one of the fast-growing areas of the livestock sector in India as well as in the world. It shows 5%–6% annual growth rate in eggs and 7%–8% annual growth rate in broiler meat production in India. The 20th livestock census shows a sharp increase in backyard poultry birds between 2012 and 2019 which was 46% (317.07 million), as compared to commercial poultry farms which increases to 4.5% (534.74 million) and the total poultry in the country is 851.81 million in 2019, increased by 16.8% over the previous census of 2012. These include fowls, ducks, emu, turkeys, quail, and other birds. India is the third-largest egg producer

Concepts of Advanced Zero Waste Tools
https://doi.org/10.1016/B978-0-12-822183-9.00005-2

Fig. 1 Poultry farm.

of the world and annually produces around 75 billion eggs, the majority of it comes from commercial farms. The first and second places occupied by China and the USA, respectively (20th Livestock Census, 2019). Crop production alone may not solve the food requirement of the fast-growing population of the country (Karupannan et al., 2020; Dowlath et al., 2020). The poultry production has been the appropriate answer for the fulfillment of requirements of a balanced diet. In India, poultry meat contributes 47.05% of total meat production (Olexa and Goldfarb, 2008). Over 66 billion chickens are slaughtered for meat in the world each year. Total global chicken meat production is nearly 110 million tons per annum. The largest producer of chicken by weight is the USA, accounting for nearly 18% of world's chicken meat production. The highest producer of chicken by numbers, however, is China, accounting for over 14% of world production of broiler chickens. India is the fifth-largest chicken meat producer in the world, which produces 3.5 million tons of broiler meat, whereas the USA, Brazil, China, and Russia stand first four respectively (FAO, 2017). Nearly 20 million people are employed in the poultry industry with around 1000 hatcheries operating across India (Annual Report of Department of Animal Husbandry, Dairying and Fisheries, 2015).

The poultry industry produces large amounts of solid waste, such as excreta, feed, feathers, hatchery waste, mortality waste, and so on, which is disposed of by burial, incineration, rendering, or landfilling (Blake, 2004). Each of these processes, however, has certain drawbacks such as cost involvement, labor intensiveness, production of environmental pollutants and odor, and so on (Hussain and Kecili, 2019; Hussain, 2019; Kumar et al., 2007). Inadequate approach and carelessness of disposal of poultry slaughterhouse waste products will lead to the constant threat of disease ailments on poultry farms. This results in direct losses in the form of mortality and reduced productivity. The appropriate utilization of its waste or by-products increases the monetary output and protects from its unwanted side effects (Hussain et al., 2020). Therefore, early disposal of wastes with a well-organized method is an important poultry waste management tool for raising healthy and profitable poultry farming activity. There are various methods for managing poultry waste such as land application of litter as an organic fertilizer, biogas production, products for commercial purposes, and so on (Singh et al., 2018). Therefore, developing a technically feasible and economically viable method for this purpose would benefit both large- and small-scale poultry farms and processing units (Hutchinson and Seekins, 2008).

There are many organizations involved in this global industry, representing a research network dealing with issues such as breeding, nutrition, welfare, husbandry, production, processing, product development, physiology, product quality, and economics (Ratwan et al., 2018). This chapter discusses the wastes generated by the poultry industry and the methods and technologies used in the efficient handling of the wastes with a zero-waste approach.

2 Poultry processing steps and wastes generated

The steps involved in processing the poultry meat are summarized in Fig. 2, which starts with live birds and moves till meatpacking and storing.

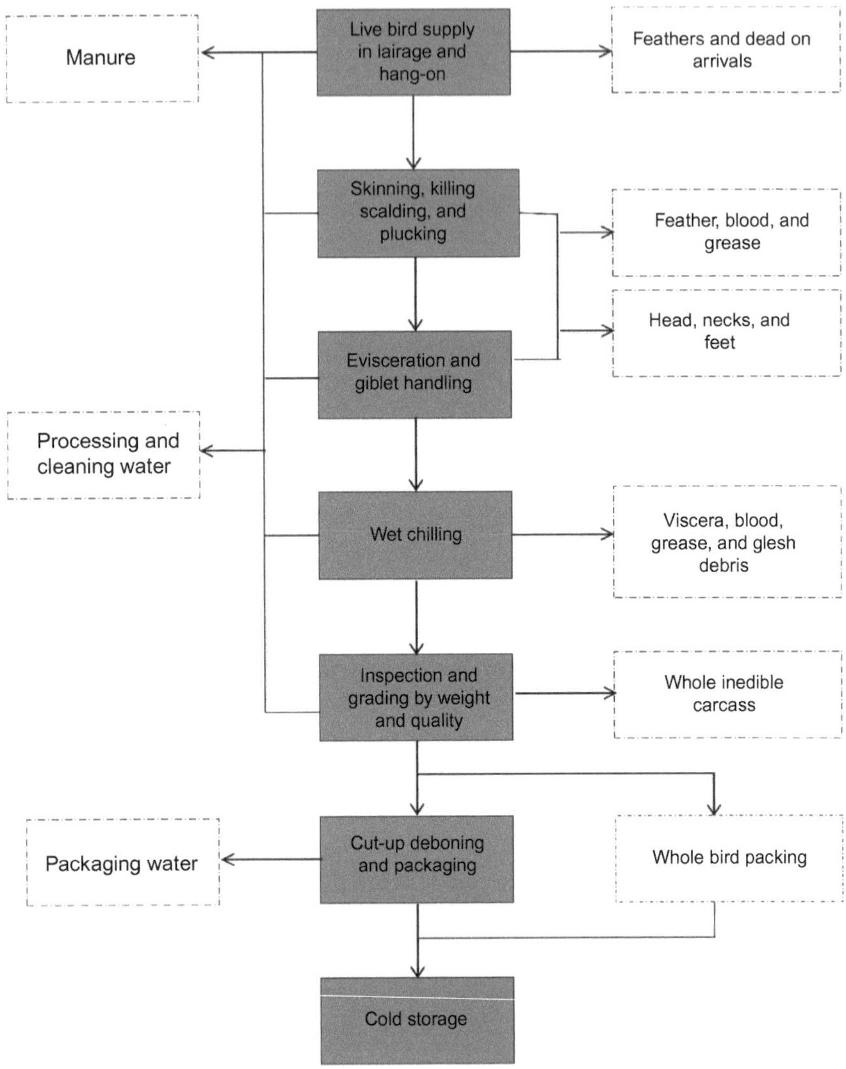

Fig. 2 Processing of poultry meat.

After the birds are killed, feathers are detached by mechanical pluckers fitted with rubber fingers on rotating discs, and the process is finished by operators called pinners, who manually finish the process of plucking. Detached feathers are forced out over a screen into a vessel that holds the eluted blood, cleaning water, and feathers mixture.

Poultry industry contributes to a variety of waste in the process of poultry meat processing. This includes feathers, skin, blood, heads, legs, and internal

Fig. 3 Nonedible poultry wastes.

organs like the liver, lungs, kidneys, brain, spleen, and so on (Fig. 3). The nonedible by-products are mostly used for the manufacture of chicken meal. The poultry industry is not just slaughtering of birds and processing it to reach the customer but also involves the rearing of birds up to the stage of slaughtering. During the rearing of birds, there is a consistent amount of waste being generated as in the form of bedding material, a dead bird, spent birds, eggshells, excreta, feathers, and so on. It is indeed necessary to concentrate on managing the waste generated during the growth of poultry birds. Improper disposal of poultry wastes causes infectious diseases and reduces the production process of poultry. Proper disposal and strong management procedures of these wastes help to maintain profitable and healthy farms. Improper disposal of these wastes creates serious environmental problems (Anupoju et al., 2015).

3 Zero-waste approach in handling poultry waste

All the above-mentioned wastes have good nutritive values. These wastes can be further utilized for the production of value-added products which is a step toward zero-waste products from the poultry industry (Table 1). The widely followed two utilization methods of waste are either as animal feed or as fertilizer (Jayathilakan et al., 2012).

3.1 Feather waste

Globally, millions of tons of feather waste are generated by the poultry industry (Sinkiewicz et al., 2017). It is one of the sources of pollution

Table 1 Poultry waste utilized for the production of value-added products.

S. no.	By-products	Uses
1	Poultry litter and manure	Recycled feed, surface dressing of agricultural land
2	Hatchery wastes (eggshells, unhatched eggs, infertile eggs)	Feed, bioremediation, and fertilizer
3	Feathers	Feather meal, bedding material, and manure
4	Head	Poultry meal
5	Blood	Blood meal
6	Gizzards and proventriculus	Edible and source for chitinolytic enzyme
7	Feet	Soup, poultry grasses
8	Intestine and glands	Meat meal, sportgats, poultry grease, and active principles

because of poor management and its recalcitrant nature. Waste feathers are habitant for pathogenic micro-organisms such as Salmonella and Vibrio and emit various pollutants such as ammonia, nitrous oxide, and hydrogen sulfide which harmfully affect human health and the environment. Chicken contains approximately 5%–7% of feathers in their total body weight. Feathers mainly consist of keratin protein. It is insoluble in water, weak acids, and organic solvents, and it is present in nails, wool, scales, and so on. Difficult to degrade by using common proteolytic enzymes (trypsin, pepsin) due to the cross-linking hydrogen bonds, salt linkages, and disulfide linkages. Feather keratin consists of alpha-helical and beta-sheet conformation. Outer quill is made up of beta-sheets. The beta-sheets are harder than alpha helix due to the presence of high cysteine content. Cysteine contains disulfide bonds that link adjacent keratin subunits. Other than cysteine, amino acids like lysine, proline, valine, threonine, and glutamine are predominantly present in feathers. Glycine, leucine, methionine, isoleucine, histidine, glutamic acid, phenylalanine are also present in negligible amounts (Peng et al., 2019). Feathers are a source of peptides and minerals. The feather keratin is hydrolyzed by three methods, and it has a wide range of applications (Fig. 4).

1. Physical treatment
2. Chemical treatment
3. Biological treatment

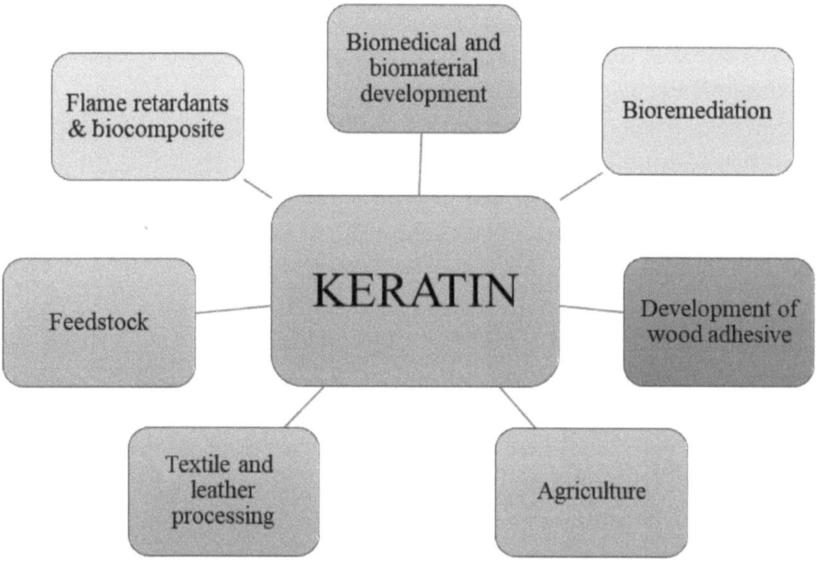

Fig. 4 Applications of feather keratin.

3.1.1 Physical treatment

Commonly, physical treatment is divided into two types, namely, mechanical treatment and thermal treatment. During mechanical treatment, to decrease the size of the particle substrate, the chicken feathers are milled and ground. The milling and grinding of feathers help in easing the process of feather hydrolysis in the subsequent treatment. However, as this method is known for huge amounts of energy consumption (Muhammad Nasir and Mohd Ghazi, 2015) and expensiveness, it is not very much practiced for conducting large-scale process. Another mechanical method is the use of ultrasounds for breaking and destructing the feather biomass. The proficiency of this method is affected by factors like frequency, duration, and the properties of the substrate (Teghammar, 2013). However, thermal treatments, such as autoclaving, steam heating, and pressure cooking, are some methods used to break the structure of proteins present in the feathers. Thus, the solubility of the feather is increased which improves the process of feather hydrolysis (Hii et al., 2014). Again, these methods consume enormous energy. Other treatments like irradiation using microwaves are also carried out for the disruption. In this treatment, the heat produced is penetrated through the substrate to its interior, thereby facilitating the disruption of cell structure (Yang et al., 2015). This method is fast and can be easily handled (Kratchanova et al., 2004).

3.1.2 Chemical treatment

In the chemical treatment, to extract soluble keratins from the chicken feathers, strong acids or alkalis are used to cleave the disulfide bonds (Zhao et al., 2012). Alkali or acid hydrolysis is nonspecific. Hypothetically, these chemicals are known to attack the peptide bonds present in the feather and break it down as short chains of peptides or aminoacids (Al-Bahri et al., 2009). Parvé (1956) suggests that there is an advantage of using acid hydrolysis over alkali hydrolysis that it destroys the amino acid tryptophan and partially destroys serine, threonine, and cysteine but does not destroy the optical activity of the amino acid. The asparagine and glutamine are converted into aspartic acid and glutamic acid, respectively. In a study, chicken feather hydrolysis was achieved to more than 85% using KOH (0.6%) for 24h at 70°C, which produced about 326mg/mL of free amino acids. However, the hydrolysis with acid at similar parameters is not efficient (Stiborova et al., 2016). In another study, feather degradation of about 78% was achieved using 1.0M NaOH hydrolysis for 24h at room temperature. Though the alkaline treatment is efficient in hydrolysis but can affect the quality of protein obtained (Kim et al., 2001). Oxidizing agents like permanganate, bromine, hydrogen oxide, and so on and reducing agents like sodium sulfide, thioglycolate, potassium cyanide have also been used in hydrolysis for breaking the disulfide bonds (Forgács, 2012). Feather hydrolysis using reducing agents was carried out at pH 10–13 for 6h at 30°C, and soluble proteins were obtained. Feathers are treated with calcium hydroxide to obtain the liquid product (rich with amino acids and polypeptides). Fifty-four percentage of calcium obtained from this solution by carbonating. It can be used as an animal feed, which has a similar nutritional value of cottonseed and soybean meal. Feathers are disposed of commonly by incinerating and chemical hydrolysis methods. These methods require a large amount of energy, which reduces the quality of proteins and essential amino acids (lysine, methionine, and tryptophan). The treatment forms nonnutritive amino acids such as lysinoalanine and lanthionine (Zhao et al., 2012).

3.1.3 Biological treatment

Owing to some disadvantages associated with physical and chemical treatments, such as the usage of harsh chemicals, increased energy consumption, and added investments, alternative treatments, are being explored. To overcome these problems, biological treatments are adopted for the degradation of feathers. Biological treatment is classified into two types, namely, microbial treatment and enzymatic treatment. As the major protein present in the feathers are keratins, microbes producing keratinase enzymes are studied for the hydrolysis of feathers. Microbes convert feather into feather lysate, and the digestible protein is used for feed. The microbes secrete the keratinase enzyme. The proteinase enzyme hydrolyzes the feather keratin, collagen, and elastin. This process converts the feather waste to nutritionally balanced and digestible useful feather lysate. The common occurrence in nature of microorganisms that readily and, in some cases, preferably grow on keratinaceous substrates has supported the general belief that certain microorganisms can digest keratin (Peng et al., 2019). Thus, microorganisms and enzymes have been the focus of several studies and have reported many efficient keratin degraders.

Keratinase enzymes can break the strong bonds of the keratin in feathers (Tiwary and Gupta, 2010). To date, numerous microorganisms of bacteria, fungi, and actinomyces origin with keratinolytic properties have been identified and used. Dermatophytes are one such known fungi with keratinolytic property. These dermatophytes are known to cause diseases related to skin. Owing to the associated pathogenicity, this fungus is not much used. However, another common fungus, *Aspergillus niger*, is thoroughly explored for its industrial importance in producing various enzymes. This can produce keratinase (Lopes et al., 2011) and has been used for complete degradation of chicken feathers at neutral pH and temperature of 30°C. Gram-negative bacteria like *Burkholderia* sp., *Pseudomonas* sp., and *Chryseobacteriums* sp. are known as prominent keratinolytic bacteria (Riffel and Brandelli, 2002). *Pseudomonas* sp. P5 was found efficient in the hydrolysis of feathers up to 70%–93% within 5 days with the production of 6.2 g/L of soluble proteins and 300 mg/mL of free amino acids (Stiborova et al., 2016). Many *Bacillus* species are also known to be predominant producers of keratinase enzymes. Various strains of *Bacillus*, such as *B. subtilis*, *B. pumillus*, *B. licehniformis*, *B. cereus*, and *B. amiloliquefaciens*, have been identified as keratinolytic bacteria (Tiwary and Gupta, 2010). Nagal and Jain (2010) achieved more than 78% of feather degradation using *B. cereus* KB043 and yielded about 1.2 mg/mL of soluble protein.

At the same time, biological treatment also involves the utilization of purified or semipurified keratinase enzymes for efficient feather degradation. This enzymatic treatment is affected by factors such as media constituents, pH conditions, temperature, and duration of treatment (Sivakumar et al., 2012). But most keratinase enzymes were optimally active under alkaline conditions (Gupta and Ramnani, 2006). In a study where keratinase enzyme isolated from *Brevibacillus* sp. AS-S10-II mutant which is induced with ethyl methyl sulfonate (EMS) was found to hydrolyze feathers up to 82% at alkaline conditions (pH 9.0–10.0) within 2 days (Mukherjee et al., 2011). And 71% and 76.5% of feather hydrolysis using crude enzymes extracted from *Alternaria tenuissima* K2 and *Aspergillus nidulans* K7, respectively at pH 8.5 within 24h. Sometimes extreme alkalophilic keratinase enzymes with enhanced activity were also identified (Mousavi et al., 2013). In general, keratin hydrolysates can be produced by the biological treatment which yields soluble proteins with minimal loss of essential amino acids (Mokrejs et al., 2011). The advantages of this method are efficient, cost-effective, and eco-friendly when compared to other treatments such as physical and chemical treatments (Li, 2019).

3.1.4 Applications of feather keratin

In wood adhesives

Wood adhesives can be prepared using feathers. The feathers are dipped in the solution containing 1% sodium chloride, 0.5% sodium dodecyl sulfide and then heated at 70°C for 2h to obtain the keratin glue. The protein molecules are used for the attaching of wooden surfaces. The prepared plywood has good strength compared to the commercial plywood in dry conditions. While heating the feathers in alkaline solution, the unfolded protein and exposed polar functional group interact with the functional groups of wood veneers, which shows the result of appreciable adhesion (Adhikari et al., 2018).

In biomaterial development

Feather keratin is one of the commercial, inexpensive, and eco-friendly biomaterial (Shi et al., 2014). The feather keratin is small and uniform size protein, and the molecular weight is around 10kDa (Ullah et al., 2011). It consists of hydrophobic forces, hydrogen bonds, and covalent interactions such as disulfide bonds (Wang and Cao, 2012). The presence of reactive functional groups in keratin, which makes it chemically reactive under favorable conditions. After modification of keratin, it can be used to develop the nano/micro nanoparticles, gels, beads, and films. It is also used in the pharmaceutical, food sciences, green chemistry, medical, agricultural, cosmetics, and various other industries (Dominguez-Benetton et al., 2013; Khosa and Ullah, 2013). Hydrolyzed keratin has vast applications in the cosmetic industries. Keratin and combination of polymers (collagen, silk fibroin, and chitosan) blends were used in the cosmetic industries, and keratin-based cosmetics were used in the treatment of the skin and human hair. Keratin present in the skin and hair interacts with cosmetics which helps to maintain the moisture of the skin (Sionkowska, 2015). High molecular weight keratin is mostly used for skincare applications because of hydrophilic and film-forming characteristics. The smooth and soft sensation of the skin due to the formation of the film or coating of the keratin molecule.

In agriculture

Feather keratin is a good source of nitrogen and is used in the preparation of biofertilizers. Keratins of feathers are degraded by keratinolytic enzymes produced by bacteria and fungi and are used for composting (Reddy, 2015). The *Bacillus* genus and actinomycetes involve in keratin degradation by the production of plenty of keratinolytic enzymes (Tiquia, 2005; Tiquia et al., 2005). The hydrolyzed feathers were used as fertilizers to enhance the growth of the banana plant, ryegrass, root, and shoot dose of other cultivation crops (Gurav and Jadhav, 2013).

In biomedical applications

Keratin is a site for cellular attachment. Fibronectin and cellular-binding motifs are found in the feather keratin as well, which contains the cell adhesion sequences (Marshall et al., 1991). This feature of keratin made it into an ideal composite for the development of biomaterial for biomedical applications. Single units of keratin can penetrate the hair cuticles and skin and then nourish them without any side effects. Oxidization of (keratinous) material and cleaving some of the disulfide bonds lead to the formation of water-soluble proteins which are used as wound healing agents. The solid base of keratin research helps in the development of many biomaterials for biomedical applications (Rouse and Van Dyke, 2010). The feather keratin is also used to develop the two- and three-dimensional scaffolds and protein fibers for tissue engineering applications (Xu et al., 2014). Extracted keratin can self-congregate and polymerize into complex 3D structures that have its application as scaffolds (Rouse and Van Dyke, 2010). Keratin-based composites are fabricated as nanofiber by electrospinning which helps in regenerative medicines and tissue culture applications (Edwards et al., 2015; Boakye et al., 2015). Keratin film is used for the controlled drug delivery system (Yin et al., 2013). Keratin derived from feathers is one of the potential sources for regenerated fibers. These fibers have similar properties of sheep wool (Xu et al., 2014). Micro and nanoparticles are prepared by using feather keratin. Sun et al. (2009) prepared micro nanoparticles having ion sorption capacity due to its hydrophilic nature. Nanoparticles prepared from feather keratin are used for the controlled release of the drug. These nanoparticles have good biocompatibility and stability compared to other nanoparticles. The nanoparticles do not require any crosslinking or chemical modifications, so it is suitable for biomedical application (Hussain and Mishra, 2018; Hussain and Kharisov, 2017; Xu et al., 2014). Graphene oxide has potential applications in biomaterials. Graphene grafted with keratin enhances the attachment of Escherichia coli graphene film (Amieva et al., 2015). Yin et al. (2013) extracted 20 kDa keratin from feather used for the formation of the film. The film is used in the controlled drug release due to its mechanical properties because of the large amount of cysteine linkage.

In feedstock

Keratin is used in the preparation of animal feed. Feather keratin after hydrolysis has the parallel composition of cotton seed and soybean protein. Feather lysate thus obtained from biological treatment is a rich source of protein, nitrogen, and minerals, which can be potentially used as animal feed. This can be used as feed supplements for cattle (Brandelli et al., 2015).

In environmental remediation

Alpha helix and hollow fibers present in the feathers build a uniform structure of microporous, which is used as electrode material due to high surface area and eco-friendly (Zhan and Wool, 2011). The keratin-based material reforming the bio-based eco-friendly material due to its biocompatibility, mechanical durability, and natural abundance (Balaji et al., 2012). Modulus and strength of a single feather are better than a wool fiber (Li and Wang, 2013). Thermoplastic films made of feathers are used for the packaging of food and other applications (Reddy et al., 2013). Feathers are used to develop commercial products by acetylation and etherification processes (Reddy et al., 2013; Shi et al., 2014).

In textile

Yang and Reddy made hand-spun yarn by blending barbs (feathers fibers) and cotton. Sizing agents are used to improve the weaving performance in textile industries. Feathers are used as sizing agents; it helps to enhance the tensile strength and abrasions resistance in textile yarns (Yang and Reddy, 2013; Reddy et al., 2014a). Polyvinyl alcohol is used as a sizing agent in textile industries, but it is costly and difficult to degrade, so feathers are a good substitute for it (Reddy et al., 2014b).

In leather processing

In leather processing, keratin hydrolysate is used in filling cum retaining operations in processing (Karthikeyan et al., 2007). Keratinase isolated from feathers is used for dehairing of skins/hides during the manufacture of leather (Balakumar et al., 2013). Keratinase containing detergents effectively removes the paint, blood, and dye stains from fabrics (Manivasagan et al., 2014).

As flame retardants

Feathers are used as flame retardants due to the presence of high contents of nitrogen. P-N based flame retardant prepared by using hydrolyzed feathers. Yellowish feather-based flame retardant formed is used in cotton fabrics (Wang et al., 2014).

As biocomposites

Biocomposites developed from feathers were used as matrix and reinforcement. The degradable biocomposites were formed by the use of chitosan-starch as matrix and feather component as reinforcement (Flores-Hernández et al., 2014). Feather fibers were mixed with polyethylene and form a thin composite sheet. Changes in the feather concentration did not affect the strength of composite films but improved the modulus barrier properties of films (Spiridon et al., 2012). The feather fibers were mixed with high-density polyethylene formed nonwoven mats. Mixing and compressing of reinforcing fiber and matrix form the composites (Huda and Yang, 2009).

In biodiesel production

Fat is extracted from the feather meal, in boiling water and subsequently transesterified by using nitrogen, potassium, and methane. A 7–11 percentage of biodiesel can be produced by the eco-friendly method. Compared to other feedstock feather meals produce good quality biodiesel, and it is confirmed by ASTM analysis (Thyagarajan et al., 2013).

3.2 Hatchery waste

Hatchery waste is a major waste in the poultry industry. It comprises the dead embryo, empty eggshells, infertile egg, dead chickens, and decaying tissues (Fig. 5). Traditionally, the wastes are disposed of by landfill, incineration, rendering, and composting. Hatchery waste contains high protein with 71% of moisture. The wastes are used to develop protein feed and organic fertilizer by various treatments. The vacuum extraction system used

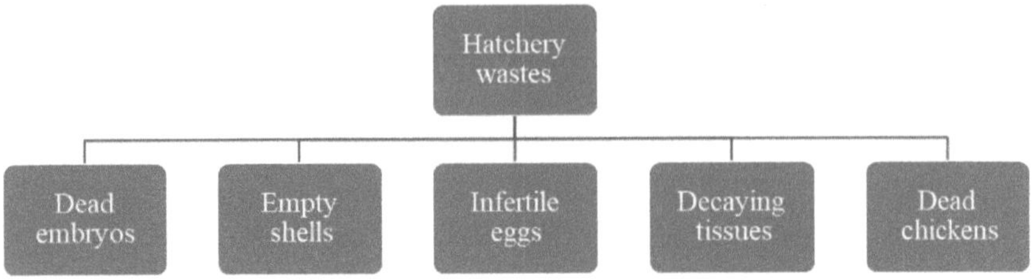

Fig. 5 Major wastes in hatchery.

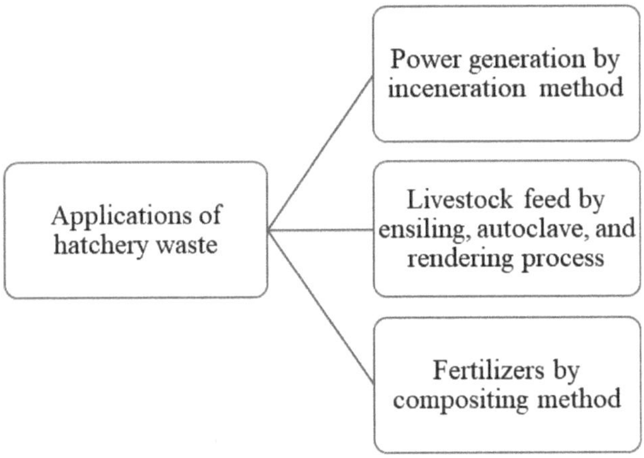

Fig. 6 Application of hatchery waste.

to transfer the wastes to the bin. Liquid and solid wastes are separated by the centrifugation process. Liquid waste is refrigerated and transferred to pet food manufacturing plants and solid wastes undergo various treatments and used for diverse applications (Fig. 6) (Glatz et al., 2011).

The hatchery wastes are used for power generation where incineration of wastes produces steam. The steam has the power to move turbine generator for the generation of electricity (Kennedy and Kennedy Leo, 1986). By rendering process, it simultaneously dries the material and separates proteins and fats. The protein and fat meals are used to feed the poultry and pig industries (Salminen and Rintala, 2002). Autoclaved hatchery wastes were dried and powdered by different methods and used as poultry feed (Said, 1996). Dead embryos were boiled at 100°C for 30 min and removed shells by soaking 20 min in cold water. The obtained product is sundried for 4 days and used as poultry and livestock feed (Abiola and Onunkwor, 2004). Ensiling is a method to preserve the nonfertile and dead embryos. The wastes were mixed with a 1:1 ratio of formic acid and propionic acid, which is then used as animal feed (Kompiang, 1994). Culled birds were treated with INSTAPRO enzyme or NaOH, the product is fermented (with added sugar) for 21 days. The fermented product is autoclaved at 124 kPa and 127°C for 90 min, then air-dried at 60°C. The advantage of this treatment is to improve the availability of essential amino acids and to make the poultry meal easily digestible by birds (Kim and Patterson, 2000). Composting of hatchery wastes are used as good organic fertilizer; it contains a high content of calcium and micronutrients. The composting is performed by (1) anaerobic digestion, (2) anaerobic co-digestion, and (3) two-stage anaerobic digestion. The unseparated hatchery wastes composed of the commercial in-vessel composter. It composts poultry waste within 4 days by minimal labor and mechanical devices. Benefits include the generation of electricity, biogas production, and products such as fertilizer and livestock feed. For anaerobic digestion, methanogen and acetogenins were used. Methane-producing bacteria are called methanogens. Acetic acid-forming bacteria are called as acetogenins. Anaerobic digestion mostly eliminates the pathogens and after 72 h digestion (Glatz et al., 2011).

3.3 Poultry litter and manure

Worldwide, the poultry industry is a large and rapidly growing agro-based industry. The poultry meat has demand because of the acceptance of most of the societies and its low cholesterol content. Poultry manure (PM) and litter cause environmental problems due to intensive production. Large-scale accumulation causes a serious environmental issue if may not apply the eco-friendly and cost-effective management technologies (Sharpley et al., 2007). Manure and litter of the poultry industry are widely used in the agricultural lands (Fig. 7). This is the viable way to recycle the Nitrogen (N), phosphorus (P), and potassium (K) available in manure (Kaiser et al., 2009). It is an organic nutrient source for crop cultivation. The excreta contain 40%–70% of nitrogen and small quantities of ammonia and urea. Storage of manure in open-air causes rapid loss of nitrogen because of its high proteolytic activity. Therefore, rapid processing preserves the nutritional properties of manure. The obtained manure when used for agricultural land increases the corn yield by altering the properties of soil, enhancing the availability of nutrients, maintaining the pH, ion exchange, organic matters, soil tilth, and water-holding capacity. The more the addition of manure, the more was the corn yield. It has been well known that poultry waste is an

Fig. 7 Applying poultry manure to agricultural lands enhances growth of plants.

organic source of valuable nutrients and micronutrients required for the rich growth of plants (Harmel et al., 2009). The use of PM in the soil helps in overcoming the waste management issues and helps in enhancing the fertility of the soil (McGrath et al., 2010). For instance, the continuous use of fertile land will cause deterioration of soil structure which leads to the reduction of crop yield. In such a case, the mixing of PM to that soil will enhance the fertility of the soil (Adeli et al., 2009).

It is common to use poultry litter as a source of nutrients for crops like cereals, forage, and fiber. When adding poultry litter to the soil for growing the grass varieties such as tall fescue, Bermuda grass, and orchard grass, it improved the production of dry matter (McGrath et al., 2010). In a study, it is proved that poultry litter is a rich source of nitrogen and metals, such as iron, copper, zinc, and manganese, which satisfies the growth requirements of plants (Tewolde et al., 2005). Poultry litter when applied in lands not only increased the yield of lint but also improved the quality of the fiber (Tewolde et al., 2009; Adeli et al., 2009). To evaluate the fertilizer value of the freshly applied and residual PM, field studies were performed by Massey University, New Zealand. Fresh PM, after composting for 5 weeks in the presence of sulfur and rock phosphate under aerobic condition, was applied to land as a source of nitrogen for the cultivation of cabbage and maize subsequently in the same land. Urea was used as a control source of nitrogen. Although the cabbage yield of urea used land was high compared to manure used land, no significant difference was found between them. In the case of maize yield, no difference was found between urea and manure used lands. Thus, the PM has a higher residual value when compared to that of urea (Mahimairaja et al., 1995).

PM increased the yield of ragi and rice when combined with farmyard manure (FYM) and inorganic fertilizers (Prasad et al., 1984). PM with super-phosphate increases the wheat yield (Amanullah et al., 2007a). The PM, farm manure, and green manure increased the rice grain yield in lowland conditions (Budhar et al., 1991). Coir pith-based PM with a recommended level of NPK increased the sorghum yield (Krishnasamy et al., 2002). A combination of PM and single super phosphate recorded the highest yield in wheat (Amanullah et al., 2007b). PM increases onion bulbs yield. PM used along with mineral fertilizer increase the yield of tomato (Giardini et al., 1992). Oikeh and Asiegbu (1993) obtained the highest yield of tomato when they applied swine and PM in the ferralitic soil of Nigeria. Nakamoto et al. (1994) reported biodigested PM slurry and commercial fertilizer used for the production of sweet corns. Composted PM increases the rice grain yield (Jayanthi, 1995). In Nigeria, significant improvement of maize grain yield in degraded soil was achieved by applying the PM and equal combination of PM and inorganic fertilizer (Amanullah et al., 2007a). In Saudi Arabia, straw and grain yield of wheat was increased by applying chicken manure (Magid et al., 1995). Ugbaja (1996), in Nigeria, found that castor seed yield increased while applying PM or swine manure. Opara and Asiegbu (1996) observed the increased eggplant fruit yield by applying the PM. A combination of rock phosphate and PM increased cowpea and maize grain yield (Akande et al., 2005) (Table 2).

Table 2 Application of poultry manure and other combinations.

Composition of manure	Application
PM + FYM + inorganic fertilizer	Increased the yield of ragi and rice
PM	Increased rice grain yield
PM + super phosphate	Increases the wheat grain yield
Farm waste + green manure + PM	Increases rice grain yield in lowland condition
Coir pith-based poultry litter + NPK	Increases sorghum yield
PM + P_2O_5 as single super phosphate	Increases maize yield
PM	Increases onion bulb yield
PM + mineral fertilizer	Increases tomato yield
PM + swine manure	Increases tomato yield in ferralitic soil
PM + commercial fertilizer	Sweet corn production
PM	High rice grain yield
PM, PM + commercial fertilizer	Increases the grain size in maize
PM + swine manure	Increase the yield of castor seed
PM	Increases the straw and grain yield in wheat
PM	Increases the fruit yield of eggplant
Rock phosphate + PM	Increase the yield of cowpea and maize

Poultry litter is used as a livestock feed in the countries of Israel and the United States. The litter contains up to 40% of nitrogen, in that up to 60% of nitrogen exists in the form of nonprotein nitrogen (NPN) (Chaudhry et al., 1997). Uric acid is a major NPN source. When it is given to the ruminants (20%) as feed, the gut microbes convert it into ammonia. Poultry litters when fed to cattle, it satisfies the calcium, phosphorous, and crude protein requirement (Crickenberger and Goode, 1977). Dried poultry waste contains excellent sources of protein, ash, phosphorus, zinc, iron, and calcium. The total protein requirement of sheep is satisfied by feeding the poultry waste and contributes substantially to the energy of the total ration (Muller, 1980). Drying the manure in sunlight decreases the offensiveness odor and kills the harmful pathogens. Different concentrations of combustible methane produced by the water flushing process using PM. Biogas is prepared by anaerobic digestion of the manure. The biogas is used as fuel for engines, electricity generation, and other energy-consuming processes. Anaerobically digested manure is used for land treatment and feed supplement (Singh et al., 2018).

Poultry litter nutritional value is higher than legume hay. Vitamin B is present in the composted litter, which is a good source for B12 vitamin.

Processed poultry litter is one of the economical and safe sources of minerals, proteins, and energy for ruminants. The total digestible value of nutrients present in the poultry litter is similar to the average quality of hay. When we feed them, it provides a major portion of energy to ruminants (Abdel-Baset and Abbas, 2010), and it contains a high content of ash and fiber. It can be a substitute for forage during the drought period (Chauhan, 1993). PM and cassava peel are used for goat, and it doesn't show depressive effects in the efficiency of feed utilization and growth rate. It can replace the groundnut cake diet used for goats. (Adegbola et al., 1986). Bayemi et al. (2001) documented feeding poultry litter to suckler cattle is a reliable substitute for cottonseed cake. Supplementation of poultry litter up to 25% rations of camel, did not cause any adverse effects and also the incorporation of poultry litter enhanced crude fiber digestion (Abdel-Baset and Abbas, 2010). Ensiling sorghum forage or molasses with poultry waste enhanced the crude protein content of silage (Waziri and Kaltungo, 2017).

The chicken excretes 20% of undigested feed, and it contains an abundant amount of synthesized soluble vitamins and 1100–1400 Kcal/kg of energy (Tuleun, 1992). In China, poultry farmers constructed battery cages on ponds, and the PM is directly fed to fish (Rangayya, 1977). In Nigeria, dead poultries that are thrown into the ponds are recycled as inputs for the production of nitrogen. Nitrogenous waste sufficiently fertilizes the pond for growth of phytoplankton which serves as a feed for fish (Adewumi et al., 2011). Tilapia is raised using commercial feed and PM in the US. No difference was observed in the growth rate by using PM (Collis and Smitherman, 1978). Poultry litter contains phosphorus, which may pose the problems of acceleration of eutrophication. Some of the countries have banned the use of animal wastes for land applications (Ojha et al., 2020).

Composting or degradation of poultry litter in aerobic conditions takes around 4–5 weeks. The obtained product is a pathogen-free, odorless, fine structure with low content of moisture. It can be used as organic fertilizer. Carbon/nitrogen and moisture content influence the composting process. For composting, process moisture content should maintain 40%–60%. The addition of zeolites, clay, coir, $CaCl_2$, $CaSO_4$, $MgCl_2$, $MgSO_4$, $Al_2(SO_4)_3$ in PM decreases the NH_3 volatilization during composting. Additions of squeezed gape fruit peel to PM increase the amount of conserved nitrogen (approximately 80%) (Kelleher et al., 2002).

Organic matter of poultry litter is degraded by the aerobic and anaerobic methods. Composting is not a suitable method for poultry litter due to the presence of moisture content and produces environmental pollutants (Ammonia). Anaerobic digestion is an economically attractive method for the degradation of livestock and poultry waste. This method not only improves sanitation around the poultry farm but also provides organic manure for crops and fish ponds. Piggery, cattle waste produces less biogas compared to poultry litter due to the presence of high content of amino acids and protein. This process reduces the pathogens and odors and also requires little land space and recovered material can be used as feed or fertilizer. Biogas production using this method is a good alternative for fossil fuel, electricity generation, and reduction in emission of Carbon dioxide. Anaerobic digestion of poultry waste occurs in distinct stages such as hydrolysis, acid fermentation, acetogenesis, and methane fermentation. The addition of surfactants, phosphorite, and adsorbents has shown enhanced the digestion process. The end product of residual solid can be used as a soil fertilizer. (Waziri and Kaltungo, 2017).

The combustion of poultry litter provides both heat and power. The modern technique involving sophisticated gas clean-up produces energy with reduced pollution. Various types of combustion facilities like mass burn combustion, fluidized bed, cyclonic, rotary kiln, and liquid and gaseous incinerators convert poultry litter or a mixture of poultry litter with domestic/industrial wastes into fuel (Anupoju et al., 2015).

3.4 Egg shell waste

Eggs are the ingredients in many of the food products. The tons of eggshells (Fig. 8) waste are generated per day. The eggshell wastes are converted to

Fig. 8 Eggshells.

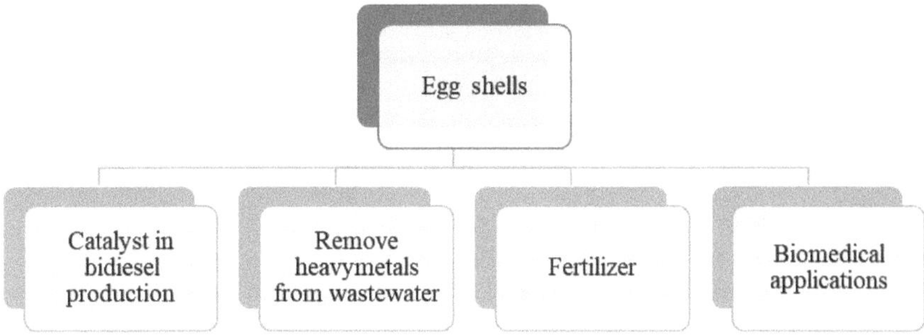

Fig. 9 Applications of eggshells.

valuable products by using current technology such as bone substitute, preparation of calcium phosphate, growth and mineralization of bone, catalyst, and used in medical and dental low-cost adsorbent (Fig. 9). Eggshell mainly consists of 94% calcium carbonate and a trace amount of magnesium carbonate, calcium phosphate, and organic matter (Ramesh et al., 2016; Eletta et al., 2016).

In general, biodiesel is produced by transesterification of vegetable or animal fats with short-chain alcohols in the presence of catalysts (Ma and Hanna, 1999). By calcining the eggshells, Wei et al. (2009) prepared solid catalyst which can also be reused. In the process of producing biodiesel, calcined eggshells used as catalysts eminently enhanced the transesterification of vegetable oil with methanol. This usage of calcined eggshells does not just reduce the contaminants by recycling but also helps in the preparation of cost-effective and eco-friendly catalysts. Thus, the cost for the production of biodiesel can also be reduced (Eletta et al., 2016).

Calcium oxide derived from eggshells used as a catalyst in palm oil, *Jatropha curcas*. Heterogeneous ($CaCO_3$) and homogenous (NaOH, KOH, and H_2SO_4) catalysts are used for biodiesel production. Owing to some disadvantages of homogenous catalyst (expensive in separation, corrosive, difficult to reuse), heterogeneous catalyst attained attention because of low cost, noncorrosive, recyclable, and simple purification and effective separation process (Laca et al., 2017). Park et al. (2007) identified that eggshells can remove heavy metals from the wastewater and as well as has a good neutralization capacity of heavy metals in strong acidic wastewaters. Eggshell was proficient in removing hydrogen sulfide and anionic dye from aqueous solution. Eggshells are used as fertilizers as it contains high levels of calcium and nitrogen source. In tomato and berry plants, eggshell fertilizers reduce the blossom end rot disease and change the pH of acidic soils and enrichment of calcium. It is also used to increase the nutritional intake of the plant. The shell contains calcium carbonate, which reduces the natural reserves of limestone. The chicken eggshells are used to make tablets that are equal to CIPCAL-500 calcium tablets. Eggshell waste calcium source was evaluated in the diet of Rhode Island Red roosters. It increases the quality of semen, gonadal development, calcium levels in plasma, and bone strength. Hydroxyapatite synthesized from eggshells creates a good source for bone treatment. The hydroxyapatite and Nano hydroxyapatite are used in the tissue regeneration and generation of bone grafts (Arabhosseini, 2018). The created grafts are nontoxic and favor of osteoblast cells. The shells are good sources of minerals that serve as pharmaceutical excipients and calcium supplements and also a component for bone implants. Minor applications of eggshells are as follows:

1. Prevents the slugs and snails
2. As a fertilizer
3. Feed for poultry birds
4. To be added in cement for increasing the strength
5. For artwork (Mosaics and 3D painting).

3.4.1 Eggshell membrane

The eggshell membrane contains glycoproteins and collagen, it can be used in cosmetic production. Membranes are separated by different techniques, which have huge medicinal applications. The membrane contains 10% of collagen. The market value per gram of purified collagen is US$ 1000. The application of eggshell membranes in medicine

1. Collagen glue used in the filling of corneal wound
2. Skin grafts
3. Angioplasty sleeves
4. Dental implants
5. Treatment of osteoporosis
6. Food casings
7. Film emulsions
8. Plastic surgery
9. Pharmaceuticals
10. Component as bioplastics
11. Alter the foodborne bacterial pathogen heat resistance using an eggshell membrane bacteriolytic enzyme.
12. Removal of reactive dyes from waste effluents
13. Removal of heavy metals in the diluted waste solution

3.5 Spent hens

Spent hens are another valuable product of the poultry industry. It is considered as a potential source of protein due to the presence of approximately 25% of dry weight as crude proteins. Mechanically deboned meat (MDM) is produced from the spent hens. MDM constitutes about 20% of live bird's weight, whereas the remaining are processed to produce animal diet supplements. The chemical treatment of the recovered proteins obtained from spent hens using protein degrading agents like urea and sodium dodecyl sulfate helped in producing wood adhesives. Emulsion-based mutton nuggets in combination with spent hen by-products such as skin, gizzards, and heart increase the acceptability (Kim and Patterson, 2000).

3.6 Offal waste

Blood and bone meals produce methane quickly. Slaughter wastes also produce high content of methane due to the presence of nitrogen content. Blood constitutes up to 7% of the live bodyweight of the chicken. It is a valuable source of heme protein and iron. Blood is considered as waste and is drained off after slaughtering the bird. This can be used for the production of both edible and nonedible products. Some of the edible products are blood meal, sausages, blood cake, biscuits, and bread, and some of the nonedible uses of blood are the binding agent, fertilizer, feedstuffs, emulsifier, stabilizer, and color additives. Blood plasma contains 60% of albumin, so it can form a gel (Silva and Silvestre, 2003). The strength of the gel is based on increasing concentration. The blood is used for laboratory media preparation (blood agar, animal tissue culture, and peptone) (Kurbanoglu and Kurbanoglu, 2004). Blood components, such as fibrinolysin, fibrinogen, kallikreins, serotonin, plasminogen, and immunoglobulins are isolated for medical/chemical uses and albumins, globulins, glycerophosphates, sphingomyelins, and catalase, are also used for biological assays. Chicken legs exist in huge amounts as a by-product of the poultry industry (Jayathilakan et al., 2012). Karthikeyan and Babu (2017) investigated the possibility of turning chicken leg skins into leather products. The raw skin was characterized and converted into finished leather by using suitable tanning methodology and the resultant chicken leg leather with attractive grain pattern has been utilized for the preparation of highly fashionable leather products. The offal wastes are pretreated and used as a feed with high contents of protein for pets (Vats et al., 2019).

4 Conclusion

The poultry meat and eggs are affordable and quality food products consumed by ethnic populations worldwide. Poultry industry produces huge amounts of wastes such as bedding material (straw, rice, peanut hulls, sawdust, wood shavings, and shredded paper), feathers, eggshells, manure, and dead birds. It is a challenging job to dispose of these wastes. One hen produces 14 kg of egg mass and 40 kg of excreta. Owing to moisture of poultry waste creates offensive odor and problems of insects. Improper handling of poultry wastes creates problems like air, water, land pollutions, and also a source for pathogenic microbes. The traditional methods of waste disposal enhanced with modern technologies help in managing poultry waste and keep a step forward to achieve zero-waste goals. Poultry waste, instead of being a source of the problem, can be used as a source of nutrients and energy. The use of technologies, such as microbial technology, bioprocessing technology, enzyme technology, and the developments in these areas, can help hand in hand for the sustainable growth of the poultry industry. The use of keratinolytic bacteria and enzymes for the eco-friendly hydrolysis of feather waste, conversion of hatchery wastes into feed and fuel by bioprocessing, use of poultry litter as a nutrient source for feeding animals, birds and fishes, conversion of various other wastes like eggshells into biomaterials and effective utilization in various fields like agriculture, medical, industrial, pharmaceutical and fuel are some of the procedures followed in the poultry industry to attain zero or near-zero waste. As farmers-owned poultry farms are in equal proportion to modern-day poultry farms, the awareness about the proper usage of the waste material, and education about the technologies must be extended to them. There is the responsibility of the policymakers too in the process of creating awareness, implementation of policies, and giving subsidies to the needy for the efficient management of waste.

Important websites

1. http://www.dahd.nic.in/about-us/divisions/statistics
2. https://thepoultrysite.com/
3. https://www.poultryworld.net/
4. https://www.wattagnet.com/

5. https://www.biomin.net/species/poultry/
6. http://www.poultryhub.org/production/husbandry-management/housing-environment/waste-management/
7. http://www.fao.org/poultry-production-products/en/
8. https://edis.ifas.ufl.edu/topic_poultry_waste_management

Reference

Abdel-Baset, N.S., Abbas, S.F., 2010. Study on the use of dried poultry litter in the camel's ration. In: Veterinary Research Forum. Vol. 2(1), pp. 65–71.

Abiola, S.S., Onunkwor, E.K., 2004. Replacement value of hatchery waste meal for fish meal in layer diets. Bioresour. Technol. 95 (1), 103–106.

Adegbola, A.A., Smith, O.B., Okeudo, N.J., 1986. Responses of West African Dwarf sheep fed cassava peel and poultry manure based diets. In: Utilization of Research Results on Forage and Agricultural By-Product Materials as Animal Feed Resources in Africa, p. 357.

Adeli, A., Tewolde, H., Sistani, K.R., Rowe, D.E., 2009. Broiler litter fertilization and cropping system impacts on soil properties. Agron. J. 101 (6), 1304–1310.

Adewumi, A.A., Adewumi, I.K., Olaleye, V.F., 2011. Livestock waste-menace: fish wealth-solution. Afr. J. Environ. Sci. Technol. 5 (3), 149–154.

Adhikari, B.B., Chae, M., Bressler, D.C., 2018. Utilization of slaughterhouse waste in value-added applications: recent advances in the development of wood adhesives. Polymers 10 (2), 176.

Akande, M.O., Adedira, J.A., Oluwatoyinbo, F.I., 2005. Effects of rock phosphate amended with poultry manure on soil available P and yield of maize and cowpea. Afr. J. Biotechnol. 4 (5), 444–448.

Al-Bahri, M.B., Al-Naimi, S.A., Ahammed, S.H., 2009. Study the effect of hydrolysis variables on the production of soya protein hydrolysis. Al-Khwarizmi Eng. J. 5 (4), 25–38.

Amanullah, M.M., Somasundaram, E., Vaiyapuri, K., Sathyamoorthi, K., 2007a. Poultry manure to crops—a review. Agric. Rev. 28 (3), 216–222.

Amanullah, M.M., Vaiyapuri, K., Sathyamoorthi, K., Pazhanivelan, S., Alagesan, A., 2007b. Nutrient uptake, tuber yield of cassava (*Manihot esculenta* Crantz.) and soil fertility as influenced by organic manures. J. Agron. 6 (1), 183.

Amieva, E.J.C., Fuentes-Ramirez, R., Martinez-Hernandez, A.L., Millan-Chiu, B., Lopez-Marin, L.M., Castaño, V.M., Velasco-Santos, C., 2015. Graphene oxide and reduced graphene oxide modification with polypeptide chains from chicken feather keratin. J. Alloys Compd. 643, S137–S143.

Annual Report of Department of Animal Husbandry, Dairying and Fisheries, 2015. Ministry of Fao. www.dahd.nic.in/documents/reports.

Anupoju, G.R., Ahuja, S., Gandu, B., Sandhya, K., Kuruti, K., Yerramsetti, V.S., 2015. Biogas from poultry litter: a review on recent technological advancements. In: Advances in Bioprocess Technology. Springer, Cham, pp. 133–147.

Arabhosseini, A., 2018. Application of eggshell wastes as valuable and utilizable products: a review. Res. Agric. Eng. 64 (2), 104–114.

Balaji, S., Kumar, R., Sripriya, R., Rao, U., Mandal, A., Kakkar, P., Reddy, P.N., Sehgal, P. K., 2012. Characterization of keratin-collagen 3D scaffold for biomedical applications. Polym. Adv. Technol. 23 (3), 500–507.

Balakumar, S., Mahesh, N., Arunkumar, M., Sivakumar, R., Hemambujavalli, V., 2013. Optimization of keratinase production by keratinolytic organisms under submerged fermentation. Optimization 5 (3), 1294–1300.

Bayemi, P.H., Tiabou, E.B., Nguemdjom, A., Kamga, P., Mbanya, J., Ndi, C., Nfi, A., 2001. Effect of replacing cotton seed cake with poultry droppings on weight gain of growing cattle at Bambui, Cameroon. Trop. Anim. Health Prod. 33 (1), 49–56.
Blake, J.P., 2004. Methods and technologies for handling mortality losses. World's Poultry Sci. J. 60 (4), 489–499.
Boakye, M.A., Rijal, N.P., Adhikari, U., Bhattarai, N., 2015. Fabrication and characterization of electrospun PCL-MgO-keratin-based composite nanofibers for biomedical applications. Materials 8 (7), 4080–4095.
Brandelli, A., Sala, L., Kalil, S.J., 2015. Microbial enzymes for bioconversion of poultry waste into added-value products. Food Res. Int. 73, 3–12.
Budhar, M.N., Palaniappan, S.P., Rangasamy, A., 1991. Effect of farm waste and green manures on low land rice. Indian J. Agron. 36 (2), 251–252.
Chaudhry, S.M., Naseer, Z., Alkraidees, M.S., 1997. Nutritive evaluation of poultry waste and sudex grass silage for sheep. Asian Australas. J. Anim. Sci. 10 (1), 79–85.
Chauhan, T.R., 1993. Nutritional evaluation of urea/poultry litter enriched wheat straw based rations to adult buffaloes. Buffalo J. 9 (3), 187–194.
Collis, W.J., Smitherman, R.O., 1978. Production of tilapia hybrids with cattle manure or a commercial diet. In: Symposium on the Culture of Exotic Fishes. American Fisheries Society, Bethesda, Maryland, USA, pp. 43–54.
Crickenberger, R.G., Goode, L., 1977. Guidelines for Feeding Broiler Litter to Beef Cattle. NC Agricultural Extension Service.
Dominguez-Benetton, X., Srikanth, S., Satyawali, Y., Vanbroekhoven, K., Pant, D., 2013. Enzymatic electrosynthesis: an overview on the progress in enzyme-electrodes for the production of electricity, fuels, and chemicals. J. Microb. Biochem.Technol. 6, 2.
Dowlath, M.J.H., Karuppannan, S.K., Rajadesingu, S., Arunachalam, K.D., 2020. Major environmental issues and problems. In: Hussain, C. (Ed.), Handbook of Environmental Materials Management. Springer, Cham.
Edwards, A., Jarvis, D., Hopkins, T., Pixley, S., Bhattarai, N., 2015. Poly(ε-caprolactone)/keratin based composite nanofibers for biomedical applications. J. Biomed. Mater. Res. B Appl. Biomater. 103 (1), 21–30.
Eletta, O.A.A., Ajayi, O.A., Ogunleye, O.O., Akpan, I.C., 2016. Adsorption of cyanide from aqueous solution using calcinated eggshells: equilibrium and optimisation studies. J. Environ. Chem. Eng. 4 (1), 1367–1375.
FAO, 2017. Food and Agriculture Organization of the United Nations. http://www.fao.org/news/archive/news-bydate/2017/en/?page=3&ipp=10&tx_dynalist_pi1[par]=YToxOntzOjE6IkwiO3M6MToiMCI7fQ==.
Flores-Hernández, C.G., Colín-Cruz, A., Velasco-Santos, C., Castaño, V.M., Rivera-Armenta, J.L., Almendarez-Camarillo, A., García-Casillas, P.E., Martínez-Hernández, A.L., 2014. All green composites from fully renewable biopolymers: chitosan-starch reinforced with keratin from feathers. Polymers 6 (3), 686–705.
Forgács, G., 2012. Biogas production from citrus wastes and chicken feather: pretreatment and co-digestion. Doctoral Dissertation, Chalmers University of Technology.
Giardini, L., Pimpini, F., Borin, M., Gianquinto, G., 1992. Effects of poultry manure and mineral fertilizers on the yield of crops. J. Agric. Sci. 118 (2), 207–213.
Glatz, P., Miao, Z., Rodda, B., 2011. Handling and treatment of poultry hatchery waste: a review. Sustainability 3 (1), 216–237.
Gupta, R., Ramnani, P., 2006. Microbial keratinases and their prospective applications: an overview. Appl. Microbiol. Biotechnol. 70 (1), 21.
Gurav, R.G., Jadhav, J.P., 2013. A novel source of biofertilizer from feather biomass for banana cultivation. Environ. Sci. Pollut. Res. 20 (7), 4532–4539.
Harmel, R.D., Smith, D.R., Haney, R.L., Dozier, M., 2009. Nitrogen and phosphorus runoff from cropland and pasture fields fertilized with poultry litter. J. Soil Water Conserv. 64 (6), 400–412.

Hii, K., Baroutian, S., Parthasarathy, R., Gapes, D.J., Eshtiaghi, N., 2014. A review of wet air oxidation and thermal hydrolysis technologies in sludge treatment. Bioresour. Technol. 155, 289–299.
Huda, S., Yang, Y., 2009. Feather fiber reinforced light-weight composites with good acoustic properties. J. Polym. Environ. 17 (2), 131.
Hussain, C., Kharisov, B., 2017. Advanced Environmental Analysis: Applications of Nanomaterials. Royal Society of Chemistry.
Hussain, C.M., Kecili, R., 2019. Modern Environmental Analysis Techniques for Pollutants. Elsevier.
Hussain, C.M., Mishra, A.K., 2018. Nanotechnology in Environmental Science, 2 Volumes. vol. 1 John Wiley & Sons.
Hussain, C.M. (Ed.), 2019. Handbook of Environmental Materials Management. Springer, Cham, Switzerland.
Hussain, C.M., Solcova, O., Kacem-Boureau, M., dos Santos, E.V., Stegemann, J., Alvarez, E.T., Karima, G., Cristaldi, A., Kuppusamy, S., Castro, P.M.L., Castelo-Grande, T., 2020. The Handbook of Environmental Remediation: Classic and Modern Techniques. Royal Society of Chemistry.
Hutchinson, M., Seekins, B., 2008. Safe Disposal of Backyard Poultry Mortalities. University of Maine Cooperative Extension, pp. 21–23.
Jayanthi, C., 1995. Sustainable component linkage and resource recycling to lowland integrated farming systems. Doctoral Dissertation, M. Sc. Thesis.
Jayathilakan, K., Sultana, K., Radhakrishna, K., Bawa, A.S., 2012. Utilization of byproducts and waste materials from meat, poultry, and fish processing industries: a review. J. Food Sci. Technol. 49 (3), 278–293.
Kaiser, D.E., Mallarino, A.P., Haq, M.U., Allen, B.L., 2009. Runoff phosphorus loss immediately after poultry manure application as influenced by the application rate and tillage. J. Environ. Qual. 38 (1), 299–308.
Karthikeyan, R., Babu, N.C., 2017. An investigation on chicken leg skin for the preparation of fashionable leather and leather products. J. Am. Leather Chem. Assoc. 112 (6), 2.
Karthikeyan, R., Balaji, S., Sehgal, P., 2007. Industrial applications of keratins—a review. J. Sci. Ind. Res. 66, 710–715.
Karupannan, S.K., Dowlath, M.J.H., Arunachalam, K.D., 2020. Phytonanotechnology: challenges and future perspectives. In: Phytonanotechnology: Challenges and Prospects, p. 303.
Kelleher, B.P., Leahy, J.J., Henihan, A.M., O'dwyer, T.F., Sutton, D., Leahy, M.J., 2002. Advances in poultry litter disposal technology—a review. Bioresour. Technol. 83 (1), 27–36.
Kennedy, L.D., Kennedy Leo, D., 1986. Solid and Liquid Waste Incinerator System. U.S. Patent 4,619,210.
Khosa, M.A., Ullah, A., 2013. A sustainable role of keratin biopolymer in green chemistry: a review. J. Food Process. Beverages 1 (1), 8.
Kim, J.M., Lim, W.J., Suh, H.J., 2001. Feather-degrading *Bacillus* species from poultry waste. Process Biochem. 37 (3), 287–291.
Kim, W.K., Patterson, P.H., 2000. Recycling dead hens by enzyme or sodium hydroxide pretreatment and fermentation. Poult. Sci. 79 (6), 879–885.
Kompiang, S., 1994, July. Ensiling of hatchery waste using a mixture of formic and propionic acids. In: Sustainable Animal Production and the Environment. Proceedings of the 7th AAAP Animal Science Congress, pp. 179–180.
Kratchanova, M., Pavlova, E., Panchev, I., 2004. The effect of microwave heating of fresh orange peels on the fruit tissue and quality of extracted pectin. Carbohydr. Polym. 56 (2), 181–185.

Krishnasamy, R., Somasundram, J., Savithri, P., 2002, August. Sewage sludge-coir pith pellets: a source of organic manure. In: 17th World Congress of Soil Science, pp. 14–21.
Kumar, V.R.S., Sivakumar, K., Purushothaman, M.R., Natarajan, A., Amanullah, M.M., 2007. Chemical changes during composting of dead birds with caged layer manure. J. Appl. Sci. Res. 3 (10), 1100–1104.
Kurbanoglu, E.B., Kurbanoglu, N.I., 2004. Utilization as peptone for glycerol production of ram horn waste with a new process. Energy Convers. Manag. 45 (2), 225–234.
Laca, A., Laca, A., Díaz, M., 2017. Eggshell waste as catalyst: a review. J. Environ. Manag. 197, 351–359.
Li, Q., 2019. Progress in microbial degradation of feather waste. Front. Microbiol. 10, 2717.
Li, R., Wang, D., 2013. Preparation of regenerated wool keratin films from wool keratin-ionic liquid solutions. J. Appl. Polym. Sci. 127 (4), 2648–2653.
Lopes, F.C., Tichota, D.M., Daroit, D.J., Velho, R.V., Pereira, J.Q., Corrêa, A.P.F., Brandelli, A., 2011. Production of proteolytic enzymes by a keratin-degrading *Aspergillus niger*. Enzyme Res. 2011.
Ma, F., Hanna, M.A., 1999. Biodiesel production: a review. Bioresour. Technol. 70 (1), 1–15.
Magid, H.M.A., Abdel-Aal, S.I., Rabie, R.K., Sabrah, R.E., 1995. Chicken manure as a biofertilizer for wheat in the sandy soils of Saudi Arabia. J. Arid Environ. 29 (3), 413–420.
Mahimairaja, S., Bolan, N.S., Hedley, M.J., 1995. Agronomic effectiveness of poultry manure composts. Commun. Soil Sci. Plant Anal. 26 (11-12), 1843–1861.
Manivasagan, P., Sivakumar, K., Gnanam, S., Venkatesan, J., Kim, S.K., 2014. Production, biochemical characterization and detergents application of keratinase from the marine Actino bacterium action *Alloteichus* sp. MA-32. J. Surfactant Deterg. 17 (4), 669–682.
Marshall, R.C., Orwin, D.F.G., Gillespie, J.M., 1991. Structure and biochemistry of mammalian hard keratin. Electron Microsc. Rev. 4 (1), 47–83.
McGrath, S., Maguire, R.O., Tracy, B.F., Fike, J.H., 2010. Improving soil nutrition with poultry litter application in low-input forage systems. Agron. J. 102 (1), 48–54.
Ministry of Fisheries, Animal Husbandry and Dairying, 2019. 20th Livestock Census—2019 All India Report. Government of India, New Delhi. http://www.dahd.nic.in/about-us/divisions/statistics.
Mokrejs, P., Svoboda, P., Hrncirik, J., Janacova, D., Vasek, V., 2011. Processing poultry feathers into keratin hydrolysate through alkaline-enzymatic hydrolysis. Waste Manag. Res. 29 (3), 260–267.
Mousavi, S., Salouti, M., Shapoury, R., Heidari, Z., 2013. Optimization of keratinase production for feather degradation by *Bacillus subtilis*. Jundishapur J. Microbiol. 6 (8), 1G.
Muhammad Nasir, I., Mohd Ghazi, T.I., 2015. Pretreatment of lignocellulosic biomass from animal manure as a means of enhancing biogas production. Eng. Life Sci. 15 (7), 733–742.
Mukherjee, A.K., Rai, S.K., Bordoloi, N.K., 2011. Biodegradation of waste chicken-feathers by an alkaline β-keratinase (Mukartinase) purified from a mutant *Brevibacillus* sp. strain AS-S10-II. Int. Biodeterior. Biodegradation 65 (8), 1229–1237.
Muller, Z.O., 1980. Feed from animal wastes: state of knowledge. FAO Anim. Prod. Health Pap. 18.
Nagal, S., Jain, P.C., 2010. Production of feather hydrolysate by Elizabethkingia meningoseptica KB042 (MTCC 8360) in submerged fermentation. Indian J. Microbiol. 50 (1), 41–45.
Nakamoto, S.T., Leung, P., Wanitprapha, K., 1994. The value of diluted poultry slurry. Bioresour. Technol. 48 (1), 25–30.
Oikeh, S.O., Asiegbu, J.E., 1993. Growth and yield responses of tomatoes to sources and rates of organic manures in ferralitic soils. Bioresour. Technol. 45 (1), 21–25.

Ojha, S., Gaikwad, S.T., Suthar, T., Gavane, A., 2020. Microbial bioconversion of poultry waste: value added products. Ind. J. Pure Appl. Biosci. 8 (1), 165–173.
Olexa, M.T., Goldfarb, I., 2008. Hazardous Waste Regulation: Biological and Animal Waste Disposal. Food and Resource Economics Department, Florida Cooperative Extension Service, Institute of Food and Agricultural Sciences, University of Florida, Gainesville, FL.
Opara, C.N., Asiegbu, J.E., 1996. Nutrient content of poultry manures and the optimum rate for eggplant fruit yield in a weathered tropical Ultisol. Biol. Agric. Hortic. 13 (4), 341–350.
Park, H.J., Jeong, S.W., Yang, J.K., Kim, B.G., Lee, S.M., 2007. Removal of heavy metals using waste eggshell. J. Environ. Sci. 19 (12), 1436–1441.
Parvé, E.S., 1956. Biochemistry, An Introductory Textbook: by F. Haurowitz. John Wiley and Sons, Inc., New York.
Peng, Z., Mao, X., Zhang, J., Du, G., Chen, J., 2019. Effective biodegradation of chicken feather waste by co-cultivation of keratinase producing strains. Microbial Cell Factor. 18 (1), 84.
Prasad, B., Singh, A.P., Sinha, M.K., 1984. Effect of poultry manure as a source of zinc, iron and as complexing agent on zinc and iron availability and crop yield in calcareous soil. J. Indian Soc. Soil Sci. 32 (3), 519–521.
Ramesh, S., Natasha, A.N., Tan, C.Y., Bang, L.T., Ching, C.Y., Chandran, H., 2016. Direct conversion of eggshell to hydroxyapatite ceramic by a sintering method. Ceram. Int. 42 (6), 7824–7829.
Rangayya, V., 1977. Poultry cum fish farm. Asian Livestock 3 (3).
Ratwan, P., Kumar, M., Joshi, P., Sharma, N., Devi, P., Deswal, S., 2018. Interventions and package of practices needed to enhance farmer's income through livestock. Res. Rev. J. Dairy Sci. Technol. 6 (2), 20–25.
Reddy, N., 2015. Non-food industrial applications of poultry feathers. Waste Manag. 45, 91–107.
Reddy, N., Chen, L., Zhang, Y., Yang, Y., 2014a. Reducing environmental pollution of the textile industry using keratin as alternative sizing agent to poly(vinyl alcohol). J. Clean. Prod. 65, 561–567.
Reddy, N., Jiang, Q., Jin, E., Shi, Z., Hou, X., Yang, Y., 2013. Bio-thermoplastics from grafted chicken feathers for potential biomedical applications. Colloids Surf. B: Biointerfaces 110, 51–58.
Reddy, N., Shi, Z., Temme, L., Xu, H., Xu, L., Hou, X., Yang, Y., 2014b. Development and characterization of thermoplastic films from sorghum distillers dried grains grafted with various methacrylates. J. Agric. Food Chem. 62 (11), 2406–2411.
Riffel, A., Brandelli, A., 2002. Isolation and characterization of a feather-degrading bacterium from the poultry processing industry. J. Ind. Microbiol. Biotechnol. 29 (5), 255–258.
Rouse, J.G., Van Dyke, M.E., 2010. A review of keratin-based biomaterials for biomedical applications. Materials 3 (2), 999–1014.
Said, N.W., 1996. Extrusion of alternative ingredients: an environmental and a nutritional solution. J. Appl. Poult. Res. 5 (4), 395–407.
Salminen, E., Rintala, J., 2002. Anaerobic digestion of organic solid poultry slaughterhouse waste—a review. Bioresour. Technol. 83 (1), 13–26.
Sharpley, A.N., Herron, S., Daniel, T., 2007. Overcoming the challenges of phosphorus-based management in poultry farming. J. Soil Water Conserv. 62 (6), 375–389.
Shi, Z., Reddy, N., Hou, X., Yang, Y., 2014. Tensile properties of thermoplastic feather films grafted with different methacrylates. ACS Sustain. Chem. Eng. 2 (7), 1849–1856.
Silva, V.D., Silvestre, M.P., 2003. Functional properties of bovine blood plasma intended for use as a functional ingredient in human food. LWT-Food Sci. Technol. 36 (7), 709–718.

Singh, P., Mondal, T., Sharma, R., Mahalakshmi, N., Gupta, M., 2018. Poultry waste management. Int. J. Curr. Microbiol. Appl. Sci. 7 (8), 701–712.
Sinkiewicz, I., Śliwińska, A., Staroszczyk, H., Kołodziejska, I., 2017. Alternative methods of preparation of soluble keratin from chicken feathers. Waste Biomass Valorizat. 8 (4), 1043–1048.
Sionkowska, A., 2015. The potential of polymers from natural sources as components of the blends for biomedical and cosmetic applications. Pure Appl. Chem. 87 (11-12), 1075–1084.
Sivakumar, T., Shankar, T., Vijayabaskar, P., Ramasubramanian, V., 2012. Optimization for keratinase enzyme production using *Bacillus thuringiensis* TS2. Acad. J. Plant Sci. 5 (3), 102–109.
Spiridon, I., Paduraru, O.M., Rudowski, M., Kozlowski, M., Darie, R.N., 2012. Assessment of changes due to accelerated weathering of low-density polyethylene/feather composites. Ind. Eng. Chem. Res. 51 (21), 7279–7286.
Stiborova, H., Branska, B., Vesela, T., Lovecka, P., Stranska, M., Hajslova, J., Jiru, M., Patakova, P., Demnerova, K., 2016. Transformation of raw feather waste into digestible peptides and amino acids. J. Chem. Technol. Biotechnol. 91 (6), 1629–1637.
Sun, P., Liu, Z.T., Liu, Z.W., 2009. Particles from bird feather: a novel application of an ionic liquid and waste resource. J. Hazard. Mater. 170 (2-3), 786–790.
Teghammar, A., 2013. Biogas production from lignocelluloses: pretreatment, substrate characterization, co-digestion, and economic evaluation. Doctoral dissertation, Chalmers Tekniska Högskola.
Tewolde, H., Armstrong, S., Way, T.R., Rowe, D.E., Sistani, K.R., 2009. Cotton response to poultry litter applied by subsurface banding relative to surface broadcasting. Soil Sci. Soc. Am. J. 73 (2), 384–389.
Tewolde, H., Sistani, K.R., Rowe, D.E., 2005. Broiler litter as a sole nutrient source for cotton: nitrogen, phosphorus, potassium, calcium, and magnesium concentrations in plant parts. J. Plant Nutr. 28 (4), 605–619.
Thyagarajan, D., Barathi, M., Sakthivadivu, R., 2013. Scope of poultry waste utilization. IOSR J. Agric.Vet. Sci. 6 (5), 29–35.
Tiquia, S.M., 2005. Microbiological parameters as indicators of compost maturity. J. Appl. Microbiol. 99 (4), 816–828.
Tiquia, S.M., Ichida, J.M., Keener, H.M., Elwell, D.L., Burtt, E.H., Michel, F.C., 2005. Bacterial community profiles on feathers during composting as determined by terminal restriction fragment length polymorphism analysis of 16S rDNA genes. Appl. Microbiol. Biotechnol. 67 (3), 412–419.
Tiwary, E., Gupta, R., 2010. Medium optimization for a novel 58 kDa dimeric keratinase from *Bacillus licheniformis* ER-15: biochemical characterization and application in feather degradation and dehairing of hides. Bioresour. Technol. 101 (15), 6103–6110.
Tuleun, C.D., 1992. The utilization of heat-treated poultry manure in chicks diets. In: Annual Conference of the NSA PAbuj a 23rd–27th March, vol. 20.
Ugbaja, R.A.E., 1996. Growth and responses of castor oil plant to sources and rates of organic manures in ferralitic soils. Biol. Agric. Hortic. 13 (3), 291–299.
Ullah, A., Vasanthan, T., Bressler, D., Elias, A.L., Wu, J., 2011. Bioplastics from feather quill. Biomacromolecules 12 (10), 3826–3832.
Vats, N., Khan, A.A., Ahmad, K., 2019. Anaerobic co-digestion of thermal pre-treated sugarcane bagasse using poultry waste. J. Environ. Chem. Eng. 7 (5), 103323.
Wang, X., Lu, C., Chen, C., 2014. Effect of chicken feather protein based flame retardant on flame retarding performance of cotton fabric. J. Appl. Polym. Sci. 131 (15).
Wang, Y.X., Cao, X.J., 2012. Extracting keratin from chicken feathers by using a hydrophobic ionic liquid. Process Biochem. 47 (5), 896–899.

Waziri, M.I., Kaltungo, B.Y., 2017. Poultry Litter Selection, Management and Utilization in the Tropics. Poultry Science, p. 191.
Wei, Z., Xu, C., Li, B., 2009. Application of waste eggshell as low-cost solid catalyst for biodiesel production. Bioresour. Technol. 100 (11), 2883–2885.
Xu, H., Shi, Z., Reddy, N., Yang, Y., 2014. Intrinsically water-stable keratin nanoparticles and their in vivo biodistribution for targeted delivery. J. Agric. Food Chem. 62 (37), 9145–9150.
Yang, J., Andre, P., Ye, L., Yang, Y.Z., 2015. The Hedgehog signalling pathway in bone formation. Int. J. Oral Sci. 7 (2), 73–79.
Yang, Y., Reddy, N., 2013. Potential of using plant proteins and chicken feathers for cotton warp sizing. Cellulose 20 (4), 2163–2174.
Yin, X.C., Li, F.Y., He, Y.F., Wang, Y., Wang, R.M., 2013. Study on effective extraction of chicken feather keratins and their films for controlling drug release. Biomater. Sci. 1 (5), 528–536.
Zhan, M., Wool, R.P., 2011. Mechanical properties of chicken feather fibers. Polym. Compos. 32 (6), 937–944.
Zhao, W., Yang, R., Zhang, Y., Wu, L., 2012. Sustainable and practical utilization of feather keratin by an innovative physicochemical pretreatment: high density steam flash-explosion. Green Chem. 14 (12), 3352–3360.

CHAPTER SIX

Zero waste hierarchy for sustainable development

Sunpreet Singh[a] and Chaudhery Mustansar Hussain[b]
[a]Mechanical Engineering, National University of Singapore, Singapore, Singapore
[b]Department of Chemistry & Environmental Science, New Jersey Institute of Technology, Newark, NJ, United States

Contents

1 Zero Waste concept: From literature's viewpoint 123
2 ZW hierarchies: Sustainable behavior 127
3 Case studies: Success stories of ZW 135
3.1 Life cycle assessment of urban waste management: A case of Italy 135
3.2 Waste management performance: Scenario of Adelaide, Australia 136
3.3 ZW trends of Africa 137
3.4 ZW in Taiwan 138
4 Conclusions 138
References 139
Further reading 142

1 Zero Waste concept: From literature's viewpoint

The term *Zero Waste* (ZW), for decades and researchers, has been traced back to its inception in 1973 when a company, *Zero Waste Systems*, was founded in America by a chemist Paul Palmer (Krausz, 2012). The primary aim of this company was to encourage the reusability of the chemical by-products, which otherwise were disposed of as waste. By the end of the 20th century, this term had gained huge popularity with the emerging environmental and sustainability movements. Despite all, the understanding and practice of the term ZW still vary, to a great extent, from waste reduction, aspirational statement, a tool for resource management, and a solution to pollution and global climate change (Russell, 2009). In the recent past, ZW has been seen as a strategy for better industrial design and waste management and is replacing the dominant waste disposal practices of landfilling and incineration (Connett and Sheehan, 2001). ZW, as a new paradigm,

Concepts of Advanced Zero Waste Tools
https://doi.org/10.1016/B978-0-12-822183-9.00006-4

promoted recapturing resources from the general waste stream, reducing consumption, and applying a life-cycle approach to product design (Dinshaw et al., 2006; Hussain and Kharisov, 2017; Hussain and Mishra, 2019; Hussain, 2019a, b; Hussain, 2020; Singh et al., 2017).

The steps toward the concept ZW have gained huge attention, especially in the developed countries, to dominate the production of the waste as well as to act as a key methodology to addressing climate change (Veleva et al., 2017). It has been believed that achieving a real ZW industry/society could not be realized in practice due to the limits of thermodynamics. The idea of ZW involves a hierarchy to inculcate waste reduction methods including prevention of the waste, reuse of the waste wherever possible, recycling the waste either directly or indirectly, composting, incineration, and landfill (Song et al., 2015a; United Nations Development Program (UNDP), 2002). Circular economy (CE) is a recent concept associated with ZW an idea that natural waste can be regenerated and human systems can mirror such natural processes. Both CE and ZW concepts can indeed be merged with the ultimate aim of eliminating waste. In the case of the manufacturing sector, ZW focuses on the back end system, whereas CE incorporates the design of products and services. CE concept is also encouraged by various countries and the European Union launched an ambitious action plan for advancing the CE with concrete waste reduction goals, regulations, and measures of success (European Commission, 2015). It has been estimated that CE can provide an economic opportunity of over $1 trillion as well as significant social and environmental benefits (World Economic Forum, 2016).

The antiquated policies, such as tax labor (in place of resources), lack of effective measurement and reporting, less attention to waste management by policy designers, stakeholders compared to climate change and water, have made the transition to ZW and CE challenging due to the economic stake. The different forms of wastes are not just endangering human health and environment but also representing multiple penalties for businesses and huge costs for disposal. Furthermore, waste poses business risks related to endorsing new regulations risks from accidental exposure (Veleva et al., 2017). A lot of industrial organizations, such as General Motors (GM), Honda, P&G, Hershey, Jones Lang LaSalle, and Unilever, have, especially, understood the benefits of reducing waste disposal costs/risks, associated brand reputation, and diversified revenue through selling unwanted products/waste (Hermes, 2014). It has been reported by the Unilever that the company had saved $1.9 million in 2013 by achieving ZW (Hardcastle, 2016). Similarly, GM made $2.5 billion revenue by recycling activities (General Motors, 2015). Nowadays, many companies and municipalities are focused on recycling and waste-to-energy (WTE) methods to achieve their waste reduction goals, rather than the more environmentally preferred options of source reduction and reuse (Zaman and Lehmann, 2013a). A model of Biogen Cambridge for waste management is shown in Fig. 1.

According to Song et al. (2015b), "*A move beyond recycling is of ultimate necessity for huge unchartered territory of higher end of waste management hierarchy in order to achieve Reuse, Reduce, and Prevention.*" As per the reported scientific knowledge of the literature, the maximum published work has focused on providing guiding principles and case studies for waste reduction. However, very few studies focused to understand the role of effective measurement and

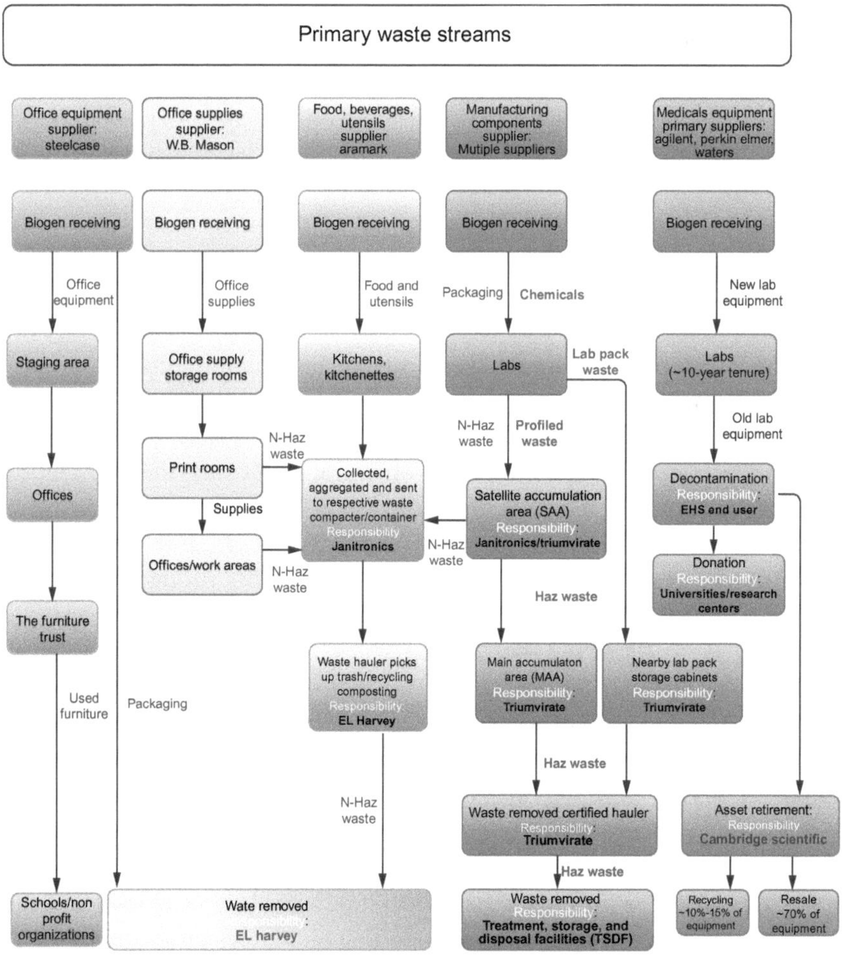

Fig. 1 A waste management model by Biogen Cambridge (Veleva et al., 2017).

employee engagement in raising awareness and advancing source reduction, reuse, and remanufacturing. A recycling council of British Columbia introduced a *6R* hierarchy for reducing waste, defined by the following: reconsider, reuse, reduce, recycle, recover, and retain. This hierarchy has been used to emphasize the importance of reduced consumption rather than recycling and recovering (Recycling Council of British Columbia (RCBC), 2014). Compared to the ZW movement, CE places a greater emphasis on using renewable stock and composting to capture and return the biological nutrients to the biosphere. China and the European Union are currently leading global efforts to enact policies, establish targets, and measure progress toward a CE (Murray et al., 2017). From ZW to landfilling goals, initiatives have been emphasized to divert from landfill as a main goal, and, to continue to rely predominantly on recycling and WTE methods of disposal (Biogen, 2015). Bartl (2014) stated, "*how do you measure something that is not there?*" In many cases, third parties collect used products, and no information is relayed back to the manufacturer or end-user of the product. The recently proposed Material Circularity Indicator also focused on the environmental aspects of material selection, design and end of life management, with no attention paid to social issues such as social justice, health, equality, and opportunities for underprivileged populations (Ellen MacArthur Foundation, 2015).

Furthermore, future research and policies to advance the CE should focus on developing goals, strategies, and guidance on minimizing simultaneously the materials (including water), energy, and waste flows at a company level while also considering related social impacts. Despite only focusing zero carbon, ZW, or water/chemical reduction, there is a need to better understand and link *"material, energy and waste process flows in a manufacturing facility from a holistic viewpoint"* (Ball et al., 2009). As similar to the application of lifecycle assessment (LCA) at the product level, it is required to develop tools to instantly indicate the minimize energy, materials (including water), and the use of hazardous chemicals at a company level (Dangelico and Pujari, 2010; Smith and Ball, 2012). Therefore, it is also important to focus on better measurement of different product design, supply chain management, and waste reduction options on greenhouse gas emissions, water, hazardous air pollution, and energy reduction.

The most widely used framework for measuring and reporting waste data currently is the *Global Reporting Initiative* (GRI) which has provided the guidelines call for reporting the total weight of hazardous and nonhazardous waste by different disposal methods including reuse, recycling, composting, recovery (including energy recovery), incineration, deep well injection, landfill, and onsite storage (Global Reporting Initiative (GRI), 2015). As an outcome, most companies report recycling, WTE, incineration, and landfill quantities. Hewlett Packard (2012), for instance, reported significant increases in reuse and recycling but the actual data provided by the company only includes pounds recycled. Hotta et al. (2016) identified four main types of recycling rate indicators based on their focus, namely, input-focused, resource recovery-focused, collection-focused, and waste diversion-focused, and called for attentiveness when using such measures. Another problem with the available indicators for measuring and reporting waste reduction practices is the declining waste as an outcome of economic recession or revenue. Many studies offered approaches for waste prevention (Gottberg et al., 2010; Sharp et al., 2010). However, currently, there is a lack of well-designed and effective indicators to help companies transition from linear to CE practices (Elia et al., 2017). Despite the business benefits of CE, developing a ZE strategy is challenging, as it requires engaging a variety of stakeholders across the supply chain, including other innovative actors as discussed by Ghisellini et al. (2016). Further, it is difficult to initially engage employees and managers as they are facing their own cost and time pressures (Du et al., 2010; Sharma et al., 2009).

Moreover, the importance of actual involvement of the contributing elements versus being informed is examined in detail by Kim et al. (2010) who analyzed the link between employees' perception of CE and ZW initiatives compared to employee participation in such activities. The researchers found that employees' participation has a direct influence and is highly effective for the enterprises to withstand with a positive relationship (Mirvis, 2012). Admittedly, through greater development, transparency, and reporting of valuable indicators can help to pass on the information and will pressurize the commercial organizations to shift from un-sustainable waste disposal to more preferred options of recycling, prevention, reduce, and reuse. However, the actual benefits could only be retained when progressive regulations, introducing new incentives for private-sector leaders, are put into action (Shahbazi et al., 2013).

2 ZW hierarchies: Sustainable behavior

Although the hierarchy of waste management varies a lot from one type of example to another, the oldest can however be traced back to the 1970s. During this period, practices were begun to consider "rubbish" for different types of treatments due to the inclusion of different categories of materials from which some were reused, recycled/composted, and buried (Schall, 1992). The concept of waste as a resource is also not new to the modern world, as the recycling or reuse of originally discarded materials, including mine waste, has been practiced since the dawn of civilization (Lottermoser, 2011). As a concept, the hierarchy makes sense in a way that is difficult to oppose. Within the context of industrial environmental management in the 1980s and 1990s, end-of-pipe responses were increasingly viewed as ineffective in their long-term impact. The essence of the different approaches was characterized by the need to avoid, eliminate, prevent, or significantly reduce the causes of environmental problems, as opposed to managing the impacts, wastes, and emissions arising further down the life cycle of the product or service (Gertsakis and Lewis, 2008). Following are the ZW legislations:

- Waste Framework Directive (2008/98/EC) (Pires et al., 2019)
- US Resource Conservation and Recovery Act (United States Environmental Protection Agency (USEPA), 2017)
- Inter-American Development Bank, definition applied at the Caribbean and Latin countries (Espinoza et al., 2010)
- Environment Protection Act (Environmental Protection Agency at South Australia (EPASA), 2018)
- Act on Waste Management at South Korea (Chung, 2011)
- Law n.12.305 in Brazil (Women in Informal Employment: Globalization and Organizing (WIEGO), 2018)

While representing the sustainable behavior of the ZW, Kumar et al., 2005 suggested that the parameters including societal support, environmental issues, product design, manufacturing, available recovery technologies, economics, suitable analysis methods, and usage patterns play crucial roles (Fig. 2). Indeed, the product's multiuse and a ZW product/material life cycle, all the indicated factors have impacts on the capability of any individual or organization to achieve the goals. To comprehend fully the significance of these factors with respect to product multiuse and ZW, the issues and barriers associated with each factor should be well described. As observed, the ZW can be established when the different factors come together for the sake of better engineering as well as utility practices. Moreover, the investors, both public and private, are also crucial in the success of ZW perspective. On a broader thought, a country or locality can achieve the

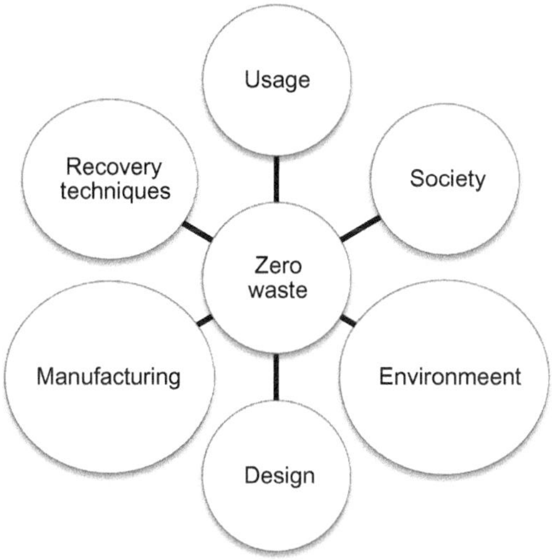

Fig. 2 Factors influencing multiuse of the products and ZW.

ZW objectives only by following the five interconnected key principles as suggested by Zaman and Lehmann (2011):

- Behavior change and sustainable consumption
- Extended producer and consumer responsibility
- 100% recycling of municipal solid waste
- Legislated zero landfill and incineration
- 100% resource recovery from waste

However, a long-term ZW vision is required as this concept is not only very hard to achieve but at the same time requires long-termed achievable initiatives. Fig. 3 shows the holistic principles for transforming any place into ZW city. Further, these ZW principles are developed based on waste hierarchy, for example, avoid, minimization, and recovery. Here, behavior changes and sustainable consumption practice will avoid unnecessary waste generation from product production and use phases. It is a fact that by achieving total recycling of waste and legislation for zero landfill and incineration, a 100% recovery of resources in the ZW city will be possible and therefore a minimum depletion of finite natural resources will be assured.

The phrase *"sustainability"* in the manufacturing industry, sometimes, offers vague information to discuss the actions related to characterizing and reducing the environmental impacts of manufacturing. However, it implies a great deal more than simply analyzing and modifying the environmental

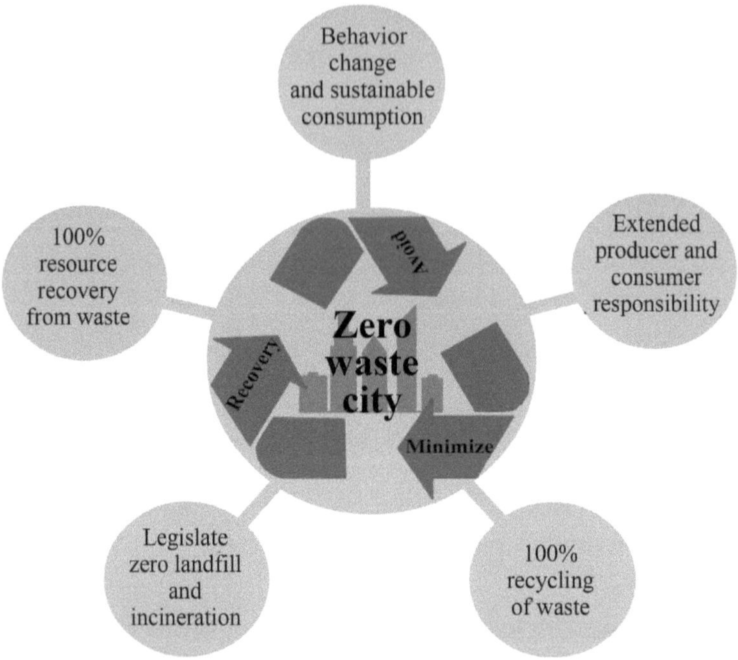

Fig. 3 Five inter-connected key principles for a ZW city (Zaman and Lehmann, 2011).

performance of manufacturing. Despite this, interpretation of sustainability should be maintained. Sustainability can only be discussed in the context of a closed system (Fig. 4) (Zaman, 2015). In the case of the manufacturing industry, sustainability is a philosophic term that cannot be considered independent of broader environmental and socio-economic systems. Further, sustainability-related impacts result from activities that manufacturing processes and systems employ to convert input materials and energy into marketable products. Similarly, according to Zaman (Zaman, 2015), the overarching policies for strategic ZW development can be broadly implemented in four phases (Fig. 5).

In the very first phase, various techniques of resource extraction, industrial ecology, cradle-to-cradle, and green engineering principles have been integrated to promote up-cycling processes by ensuring the best utilization of existing resources in society instead of extracting virgin materials. It has been recommended in this phase that the manufacturers should not fabricate products that are not easy to disassemble, recycle, and recover resources (Ziout et al., 2014). Further, avoidance, prevention, and reduction of waste are possible in these segments through sustainable design (Tseng, 2013).

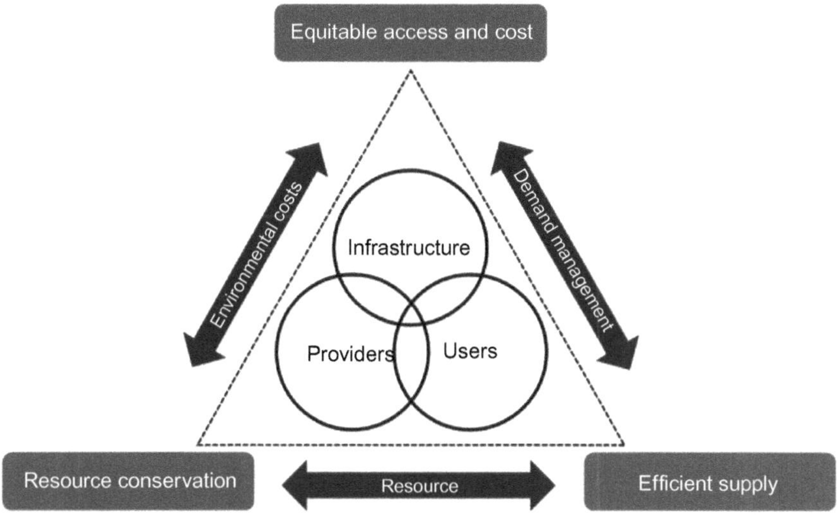

Fig. 4 Roles of sustainable model (Ziout et al., 2014).

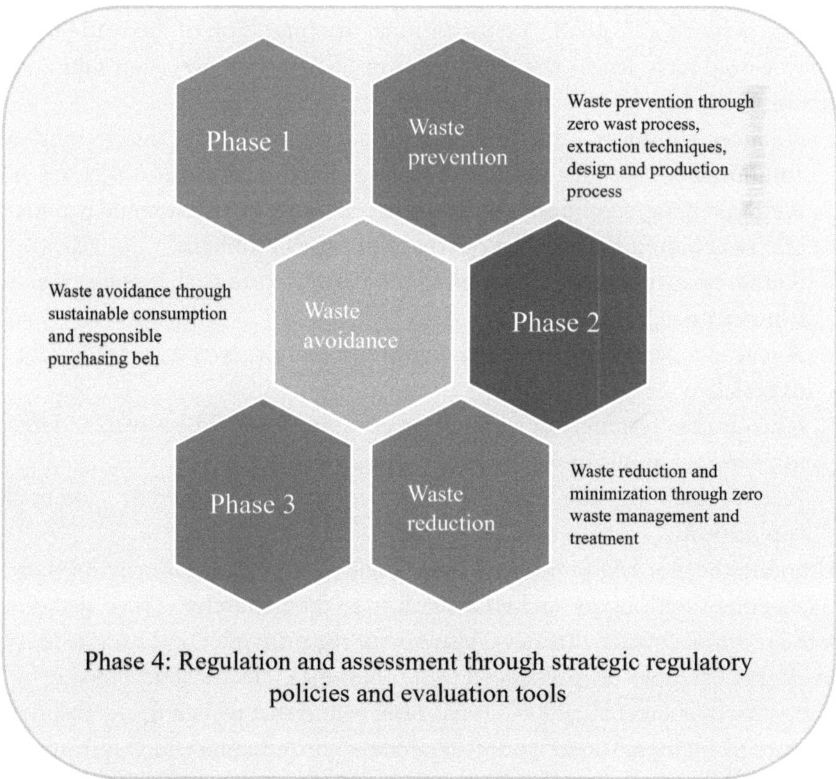

Fig. 5 Various phases for ZW development (Zaman, 2015).

Further, in the second phase, it is the responsibility of the citizens to regard the appropriate resource consumption and behavior. Indeed, sustainable consumption practices can be very important for local and global communities, as the overutilization generates an enormous amount of waste that can be minimized through responsible behavior and sustainable consumption practices (Cecere et al., 2014).

The ZW management and treatment, in the third phase, are based on "down-cycling," includes waste sorting, collection, recycling, resource recovery, and treatment followed by strict ZW guidelines to optimum resource recovery with minimal environmental degradation. ZW management should promote economic activities, by creating employment opportunities and CE society (Lee et al., 2014). In the fourth phase of ZW strategies, government regulation can play its role to guide ZW through programs, plans, and activities. The ZW policies and strategies should promote recycling activities by creating waste recycling jobs and hence contribute to circular economic growth. Therefore, holistic ZW is an approach to tackling waste problems in the 21st century, and the professionals have proposed various ideas, plans, policies, and strategies and implemented them in cities to achieve ZW goals. A conventional interpretation of the waste hierarchy would have led to the consideration of *6Rs* strategies (Gertsakis and Lewis, 2008):

- Reduce: eliminate unnecessary components or reduce the weight of components.
- Rethink: design components and the overall appliance to extend product life by avoiding faults, breakdowns, and other problems.
- Reuse/re-manufacture: a design for re-manufacture so that components from old machines.
- Recycle: a design for recycling and incorporate recycled and recyclable materials.
- Reconsider: products developed or consumed should be reconsidered for different applications, either through repair or as is.
- Retain: holding the materials or products for different potential applications.

Although the word *hierarchy* is not mentioned in the legislation, yet the waste management options are ranked according to the hierarchy. The objectives of the recent EU waste directives also mirror the principles of the waste hierarchy, for example on packaging waste, batteries, electrical and electronic waste (Rasmussen et al., 2005). It has been found that recycling and disposal of waste by incineration or landfill generate environmental effects during the treatment as recycling is associated with environmental effects from transportation, energy use, and other residuals that occur in relation to the recycling process. Incineration, landfilling, and recycling also have positive impacts on the environment because the energy produced from incineration and landfill can replace the energy produced from other sources, coal-fired electricity plants is an example. The Scottish government (Scottish Government, 2011) has introduced a waste management hierarchy (Fig. 6). It has been claimed that *"by recycling more, we can also reduce our dependence on raw materials and the associated carbon emissions for their extraction and processing."* The hierarchy starts with prevention which means to prevent the consumption of material, energy, and other resources, whenever possible. It is a fact that the consumption of any resource has an impact on the environment, either directly or indirectly.

Therefore, through proper planning, designing, and using the best available technology, consumption can be minimized. Further, it has been urged by the government to use those materials and systems which can be further reused. For instance, natural materials have the ability to easy reusability.

However, synthetic materials, like thermoplastics, are very difficult to reuse, even if these can, then only for limited applications. Recycling of the resources is also a very important pillar of the waste management hierarchy. This pillar offers the potential to treat waste materials and to convert the same into feedstock, suitable for other applications. Furthermore, the materials which cannot be reused or recycled are used for recovering energy, elements, and minerals. Finally, the in-convenient materials are finally

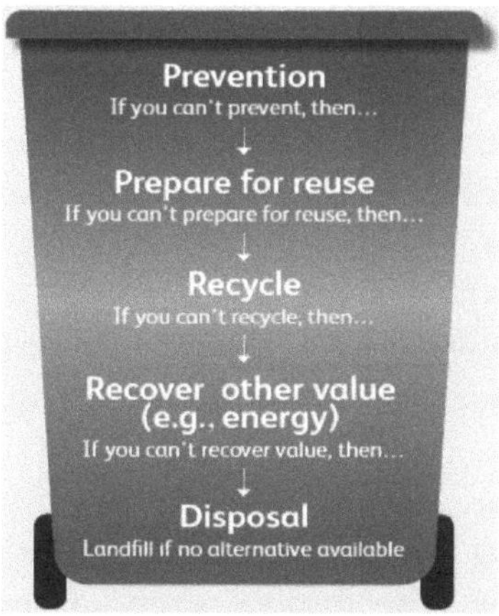

Fig. 6 A hierarchy of waste management by the Scottish Government.

disposed-off into suitable dumping plants. The maximum focus should be made on this pillar of the hierarchy as it the one posing the maximum threat to our environment. Further, this pillar of the hierarchy offers research potential where all other pillars could be strengthened. Lazarevic et al. (2010) also advocated the same for controlling plastic waste in Europe. However, they discussed waste management regimes, considering conventional principles and political objectives (Hafkamp, 2002) by following various legislations. The old statistics showed that 12 countries increased the percentage recycled by more than 10% between 2001 and 2010, and another 10 achieved increases of between 5 and 10%. Progress in enhancing recycling rates is primarily due to trends in the recycling of materials, with 19 European countries achieving fairly substantial increases in their material recycling rates since 2001 (Williams, 2015).

Various waste legislative and measures, around the globe, such as charges for waste collection and kerbside recycling collection services, have been introduced (Cole et al., 2011). The Landfill Allowance Trading Scheme resulted in waste reduction and an increase in recycling performance among the UK (Calaf-Forn et al., 2014). Greyson (2007) concluded it is ironic that the world's efforts to reduce its problems may block a preventive approach. He highlighted, "*ZW, sustainability and continued economic growth may not be achievable as they are currently practiced.*" Furthermore, education and research cannot be outweighed for their participation in achieving ZW as without proper environmental awareness and advanced research on waste, it would not be possible to achieve this goal. Also, sustainable consumption and behavior should also be considered in the ZW hierarchy as the current trend of consumption is unsustainable, it is important to understand the reality and act accordingly. Next, the ZW hierarchy is transformed industrial design, for example, eco-design or cleaner production combined with extended producer responsibility. Moreover, it is important to have specific zero depletion legislation and incentive policies as part of the strict environmental legislations (Zaman and Lehmann, 2013b). Ultimately, the refined ZW hierarchy is shown in Fig. 7.

"*A confounding 95% of the global growth over the next 40 years will happen in Asia, Africa, Latin America and the Caribbean,*" according to the Population Reference Bureau's 2009 World Population Data Sheet (Lehmann, 2011). Further, there are ways to improve waste management and change behavior in developing countries, even if there is no budget for it, for instance, Curitiba, Brazil, innovative waste collection approaches were developed, such as the "*Green Exchange Programme,*" to encourage slum

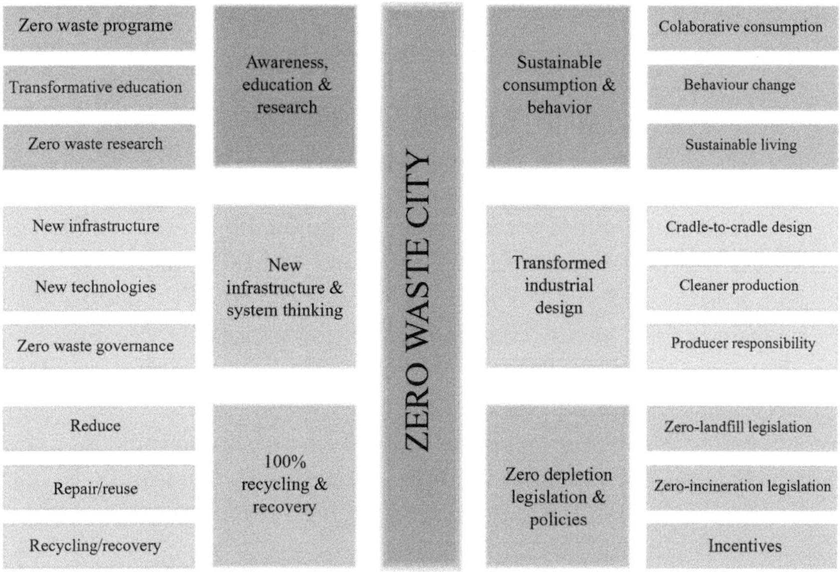

Fig. 7 Refined hierarchy for ZW (Zaman and Lehmann, 2013b).

dwellers to clean up their areas and improve public health. The other limitation existing in the ZW hierarchy can be improved by achieving dematerialization. A value-based conception of waste along with appropriate collection can prevent the loss of valuable waste. To ensure total material throughput stays within acceptable limits, waste management targets can be directly related to higher-level dematerialization goals by using material flow analysis (Van Ewijk and Stegemann, 2016).

3 Case studies: Success stories of ZW

3.1 Life cycle assessment of urban waste management: A case of Italy (Cherubini et al., 2008)

The methodological framework, life cycle assessment (LCA), used in this work has been used for the analysis of a case of municipal waste management in Rome, with an emphasis on energy and material balance (considering global and local scale airborne emissions). The emission of liquid, solid, and gaseous was carefully evaluated and classified into impact categories to calculate global warming potential (GWP), acidification potential (AP), and eutrophication potential (EP). Further, the greenhouse gases considered as potential contributors to global warming were CO_2, CH_4, N_2O, and gases responsible for rain acidification were SO_2, NOx, HCl, H_2S, and HF, and chemical compounds that contribute to aquatic system eutrophication were total-N and total-P. Four different scenarios discussed were: landfilling (S0), landfilling with biogas recovery (S1), municipal solid waste (MSW) sorting plant (S2), and direct incineration (S3). Based on LCA, the following conclusions were drawn by the research team:

- The disposal of 1 gm of waste required the production of about 0.3 gm of further waste, as abiotic matter, 0.24 gab of S0 to 0.36 gab of S3.
- The scenarios #2 and #3 able to minimize the landfill require a 20% increase of fossil energy consumption for the disposal.
- The energy synthesis through recycling of waste combustion required less environmental support than hydropower.
- Landfilling is the most polluting, and it needs to be monitored for a relatively long time period to minimize its environmental impact. S2 and S3, in spite of their benefits, showed the highest emissions.

3.2 Waste management performance: Scenario of Adelaide, Australia (Zaman, 2014)

This case study argued that achieving 100% diversion rate can be inadequate and did not reflect the core concept of the ZW philosophy. The ZW index was a new indicator used to measure and compare virgin material replacement by urban ZW management systems. Further, the ZW index quantified energy, material, and water conservation through recycling efforts rather than measuring waste diverted from landfills. Moreover, waste management performance in Adelaide during the years 2003–10 was analyzed using the proposed ZW index tool, and thereby the performance of the same during 2015–20 was predicted. The survey was designed for collecting the feedback from waste management experts in South Australia; hence, questionnaire was shared with the participants having different roles in waste management systems, including waste recycler, service provider, city council, local government, state government, nongovernment organizations (NGOs), and others. Further, Fig. 8, indicating the waste flow in Adelaide, is carefully understood.

From the analysis, the authors highlighted, "*Despite a potentially higher diversion rate of waste in 2020, Adelaide may not be significantly advanced in virgin materials substitution and resource recovery by then and the projected ZW index would be 0.45.*" Further because of the global economic crisis, the waste market has also been facing significant economic and strategic challenges in Adelaide. Hence, a strong, market-driven waste policy is urgently required.

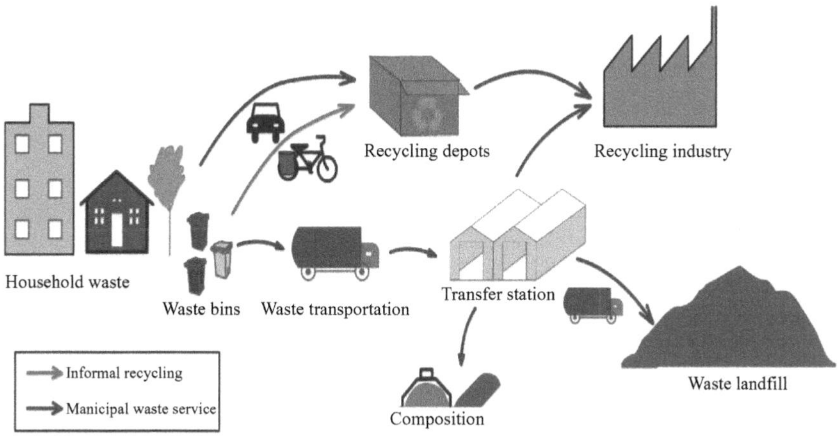

Fig. 8 Waste-flow in Adelaide (Zaman, 2014).

Adelaide lacks municipal waste data that may make forecasting waste performance faulty. Therefore, key priority areas, which are important focal points for futuristic ZW strategies in Adelaide, include capacity building, waste management rule and policy expansion, and marketplace erections.

3.3 ZW trends of Africa (Matete and Trois, 2008)

This study undertook a ZW evaluation of Africa through a designed questionnaire the attitudes of the two communities toward waste minimization and recycling, administered in Mariannhill Park and Nazareth. It has been observed that the ZW model developed for postconsumer waste in urban areas with differing levels of service led to waste minimization and recycling as necessary steps for the success of the model. Indeed, waste minimization at the point of purchase and reuse of waste within the household constituted the first step in the application of the model. Waste produced after minimization should be recycled by using at-source separation or a wet/dry model, the second step in the model. The correct application of this model, including implementation costs, neglected in this study, could lead to achieving ZW in the short term. Further discussion indicated that ZW models can be applied within existing waste management systems in South Africa; however, the success of schemes will depend on the participation rate of households. The residents of Mariannhill Park and Nazareth were willing to recycle their household waste, but their willingness to minimize waste has not been clearly established by the model.

3.4 ZW in Taiwan (Young et al., 2010)

This study was aimed to review the tasks undertaken and future plans for achieving ZW in Taiwan. For achieving this goal, waste inhibition, source decrease, rubbish to product, rubbish to energy, end-of-pipe treatment, and satisfactory disposal are the consecutive principal procedures to achieve the goal of ZW. These are regulatory alterations, consumption edification, financial inducements, technical provision, public consciousness, and chasing and writing. Stepwise targets have been set from 2005 to 2020 for MSW and industrial waste to reach the goal of ZW, with an ultimate aim to achieve 70% of MSW minimization and 85% of industrial waste minimization by 2020. To fulfill the task and to prepare an environment of ZW, the following six tasks have been implemented:

- Regulatory amendments
- Consumption education
- Financial incentives
- Technical support
- Public awareness
- Tracking and reporting

Based on the formulation, MSW has been targeted at zero landfilling ultimately, with 25% of minimization by 2007, 40% by 2011, and 70% by 2020. Further, recycling ratios for industrial waste were set at 75% by 2007, 80% by 2011, and 85% by 2020. It has been found that in 20 years, the MSW in Taiwan will be changed from 100% dumping to almost 100% properly disposed including 37% recycling. Similarly, the management of industrial waste materials has also shown an encouraging achievement with the recycling ratio to improve significantly.

4 Conclusions

This chapter highlights the ZW hierarchies as a philosophy underlying sustainable control of the different categories of waste and government policy for achieving absolute reductions in material throughput. Based on the literature reviewed, the following conclusions can be drawn:

- The ZW hierarchies are coinciding with different agendas; however, the potential for fulfillment of the set aims is still uncertain. The hierarchies are solid strategy for avoiding landfill, but uncertainty about the qualities of the hierarchy with respect to diminishing environmental hazards and natural reserve use still exists.

- Further, legislation implementation of the waste hierarchy is limited at the ground level. The limitations of the hierarchy suggest several issues, including limited powers of waste managers, leniency toward the available options, lack of attention for good rewards, lack of guidance, and decision-making, are amongst the topmost concerns.
- Available solutions that suggest improving the hierarchy systems, with respect to achieving dematerialization, conception of waste along with appropriate collection infrastructure should prevent to reduce the loss of valuable waste and increase timely reuse and recycling of expired products.
- Further, the ZW hierarchy should be strictly specified regarding open-loop and closed-loop recycling, and to ensure total material throughput within acceptable limits, the management of the waste should relate to higher-level throughput goals, wherein education and research should not be outweighed.

References

Ball, P.D., Evans, S., Levers, A., Ellison, D., 2009. Zero carbon manufacturing facility—towards integrating material, energy, and waste process flows. Proc. Inst. Mech. Eng. B J. Eng. Manuf. 223 (9), 1085–1096.

Bartl, A., 2014. Moving from recycling to waste prevention: a review of barriers and enables. Waste Manage. Res. 32 (9_suppl), 3–18.

Biogen, 2015. Corporate Citizenship Report 2014. https://www.biogen.com/content/dam/corporate/en_us/pdfs/corporate-citizenshipreport/BiogenCCR2014.pdf.

Calaf-Forn, M., Roca, J., Puig-Ventosa, I., 2014. Cap and trade schemes on waste management: a case study of the Landfill Allowance Trading Scheme (LATS) in England. Waste Manag. 34 (5), 919–928.

Cecere, G., Mancinelli, S., Mazzanti, M., 2014. Waste prevention and social preferences: the role of intrinsic and extrinsic motivations. Ecol. Econ. 107, 163–176.

Cherubini, F., Bargigli, S., Ulgiati, S., 2008. Life cycle assessment of urban waste management: energy performances and environmental impacts. The case of Rome, Italy. Waste Manag. 28 (12), 2552–2564.

Chung, S., 2011. Import and export of waste in Korea: regulation and actual practice. In: Kojima, M., Michida, E. (Eds.), Integration and Recycling in Asia: An Interim Report. Institute of Developing Economies, Chosakenkyu-Hokokushu, pp. 45–64.

Cole, C., Osmani, M., Quddus, M.A., Wheatley, A.D., Kay, K., 2011. Household waste managementin the UK: current practices and challenges. In: Castro, F., Vilarinho, C., Carvalho, J. (Eds.), Proceedings of the First International Conference on Wastes: Solutions, Treatments and Opportunities, pp. 56–61.

Connett, P., Sheehan, B., 2001. A Citizen's Agenda for Zero Waste. Grass Roots Recycling Network.

Dangelico, R.M., Pujari, D., 2010. Mainstreaming green product innovation: why and how companies integrate environmental sustainability. J. Bus. Ethics 95 (3), 471–486.

Dinshaw, A., Fortin, S., Gleason, M., Glick, D., Olivos, M., 2006. Moving Towards Zero Waste in Addison County. Middlebury College, Middlebury, VT.

Du, S., Bhattacharya, C.B., Sen, S., 2010. Maximizing business returns to corporate social responsibility (CSR): the role of CSR communication. Int. J. Manag. Rev. 12 (1), 8–19.
Elia, V., Gnoni, M.G., Tornese, F., 2017. Measuring circular economy strategies through index methods: a critical analysis. J. Clean. Prod. 142, 2741–2751.
Ellen MacArthur Foundation, 2015. Circularity Indicators e an Approach to Measuring Circularity. https://www.ellenmacarthurfoundation.org/assets/downloads/insight/CircularityIndicators_Project-Overview_May2015.pdf.
Environmental Protection Agency at South Australia (EPASA), 2018. Regulation Waste Management. https://www.sciencedirect.com/science/article/pii/S0956053X15000902.
Espinoza, P.T., Arce, E.M., Daza, D., Faure, M.S., Terraza, H., 2010. Regional Evaluation on Urban Solid Waste Management in Latin America and the Caribbean—2010 Report. Pan American Health Organization, Inter-American Association of Sanitary and Environmental Engineering, Inter-American Development Bank, Washington, DC.
European Commission, 2015. European Union: An EU Action Plan for the Circular Economy. http://eur-lex.europa.eu/legal-content/EN/TXT/?uri¼CELEX:52015DC0614.
General Motors, 2015. Waste Reduction. https://www.gm.com/content/dam/gm/en_us/english/Group3/sustainability/sustainabilitypdf/GMs_Landfill-free_Blueprint.pdf.
Gertsakis, J., Lewis, H., 2008. Sustainability and the Waste Management Hierarchy. Retrieved on January. 2003 Mar;30,.
Ghisellini, P., Cialani, C., Ulgiati, S., 2016. A review on circular economy: the expected transition to a balanced interplay of environmental and economic systems. J. Clean. Prod. 114, 11–32.
Global Reporting Initiative (GRI), 2015. G4 Sustainability Reporting Guidelines Implementation Manual. https://www.globalreporting.org/resourcelibrary/GRIG4-Part2-Implementation-Manual.pdf.
Gottberg, A., Longhurst, P.J., Cook, M.B., 2010. Exploring the potential of Product Service Systems to achieve household waste prevention on new housing developments in the UK. Waste Manag. Res. 28 (3), 228–235.
Greyson, J., 2007. An economic instrument for zero waste, economic growth and sustainability. J. Clean. Prod. 15 (13-14), 1382–1390.
Hafkamp, W., 2002. Comparison of National Solid Waste regimes in Trajectories of Change. Teoksessa: Municipal Waste Management in Europe, European Policy between Harmonisation and Subsidiarity. toim. N. Buclet, 7–26.
Hardcastle, J.L., 2016. 81% of S&P 500 Published Sustainability Reports in 2015. Environmental Leader http://www.environmentalleader.com/2016/06/30/81-of-sp-500-companies-published-sustainability-reports-in-2015.
Hermes, J., 2014. Zero Waste Strategies Create New Revenue Streams. Environmental Leader http://www.environmentalleader.com/2014/05/07/zerowaste-strategies-create-new-revenue-streams/.
Hewlett Packard, 2012. Environmental Performance. http://www.hp.com/hpinfo/global citizenship/07gcreport/productreuse/performance.html.
Hotta, Y., Visvanathan, C., Kojima, M., 2016. Recycling rate and target setting: challenges for standardized measurement. J. Mater. Cycles Waste Manage. 18 (1), 14–21.
Hussain, C.M., Keçili, R., 2019a. Modern Environmental Analysis Techniques for Pollutants, first ed. Elsevier.
Hussain, C.M., 2019b. Handbook of Environmental Materials Management, first ed. Springer International Publishing.
Hussain, C.M., 2020. The Handbook of Environmental Remediation: Classic and Modern Techniques. Royal Society of Chemistry.
Hussain, C.M., Kharisov, B., 2017. Advanced Environmental Analysis—Applications of Nanomaterials. The Royal Society of Chemistry.

Hussain, C.M., Mishra, A.K., 2019. Nanotechnology in Environmental Science. 2nd vols John Wiley & Sons.
Kim, H.R., Lee, M., Lee, H.T., Kim, N.M., 2010. Corporate social responsibility and employee-company identification. J. Bus. Ethics 95 (4), 557–569.
Krausz, R., 2012. All for naught? A critical study of zero waste to landfill initiatives (Doctoral Dissertation). Lincoln University.
Kumar, V., Bee, D.J., Shirodkar, P.S., Tumkor, S., Bettig, B.P., Sutherland, J.W., 2005. Towards sustainable "product and material flow" cycles: identifying barriers to achieving product multi-use and zero waste. In ASME 2005 International Mechanical Engineering Congress and Exposition 2005 Jan 1. American Society of Mechanical Engineers Digital Collection, pp. 433–442.
Lazarevic, D., Buclet, N., Brandt, N., 2010. The influence of the waste hierarchy in shaping European waste management: the case of plastic waste. Region. Dev. Dialog. 31 (2), 124–148.
Lee, J., Pedersen, A.B., Thomsen, M., 2014. The influence of resource strategies on childhood phthalate exposure—the role of REACH in a zero waste society. Environ. Int. 73, 312–322.
Lehmann, S., 2011. Optimizing urban material flows and waste streams in urban development through principles of zero waste and sustainable consumption. Sustainability 3 (1), 155–183.
Lottermoser, B.G., 2011. Recycling, reuse and rehabilitation of mine wastes. Elements 7 (6), 405–410.
Matete, N., Trois, C., 2008. Towards zero waste in emerging countries—a South African experience. Waste Manag. 28 (8), 1480–1492.
Mirvis, P., 2012. Employee engagement and CSR: transactional, relational, and developmental approaches. Calif. Manag. Rev. 54 (4), 93–117.
Murray, A., Skene, K., Haynes, K., 2017. The circular economy: an interdisciplinary exploration of the concept and application in a global context. J. Bus. Ethics 140 (3), 369–380.
Pires, A., Martinho, G., Rodrigues, S., Gomes, M.I., 2019. Prevention and reuse: waste hierarchy steps before waste collection. In: Sustainable Solid Waste Collection and Management. Springer, Cham, pp. 13–23.
Rasmussen, C., Vigsø, D., Ackerman, F., Porter, R., Pearce, D., Dijkgraaf, E., 2005. Rethinking the Waste Hierarchy. pp. 1–21.
Recycling Council of British Columbia (RCBC), 2014. Zero Waste Hierarchy. http://www.rcbc.ca/files/u7/RCBC%20Zero%20Waste%20Hierarchy%20-%20APR%2014.pdf.
Russell, J.K., 2009. From Solid Waste Management to Solid Waste Avoidance: A Critical Evaluation of Zero Waste Strategic Plans. State University of New York at Buffalo.
Schall, J., 1992. Does the Solid Waste Management Hierarchy Make Sense?: A Technical, Economic and Environmental Justification for the Priority of Source Reduction and Recycling. School of Forestry and Environmental Studies, Yale Univ.
Scottish Government, 2011. Policy Statement Zero Waste Regulations. http://www.gov.scot/Resource/Doc/360341/0121809.pdf.
Shahbazi, S., Kurdve, M., Bjelkemyr, M., Jönsson, C., Wiktorsson, M., 2013. Industrial waste management within manufacturing: a comparative study of tools, policies, visions and concepts. In: Shehab E., Ball P., Tjahjono B., (eds.) Proceedings of the 11th International Conference on Manufacturing Research (ICMR2013), 19th–20th September 2013, pp. 2053–3373. Cranfield University, ISBN: 978-1-907413-23-0ISSN.
Sharma, S., Sharma, J., Devi, A., 2009. Corporate social responsibility: the key role of human resource management. Bus. Intell. J. 2 (1), 205–213.
Sharp, V., Giorgi, S., Wilson, D.C., 2010. Methods to monitor and evaluate household waste prevention. Waste Manag. Res. 28 (3), 269–280.

Singh, S., Ramakrishna, S., Gupta, M.K., 2017. Towards zero waste manufacturing: a multidisciplinary review. J. Clean. Prod. 168, 1230–1243.
Smith, L., Ball, P., 2012. Steps towards sustainable manufacturing through modelling material, energy and waste flows. Int. J. Prod. Econ. 140 (1), 227–238.
Song, Q., Li, J., Zeng, X., 2015a. Minimizing the increasing solid waste through zero waste strategy. J. Clean. Prod. 104, 199–210.
Song, Q., Li, J., Zeng, X., 2015b. Minimizing the increasing solid waste through zero waste strategy. J. Clean. Prod. 104, 199–210.
Tseng, M.L., 2013. Modeling sustainable production indicators with linguistic preferences. J. Clean. Prod. 40, 46–56.
United Nations Development Program (UNDP), 2002. Results Based Management: Concepts and Methodology. https://www.epa.gov/smm/sustainable-materialsmanagement-non-hazardous-materials-and-waste-management-hierarchy.
United States Environmental Protection Agency (USEPA), 2017. Criteria for the Definition of Solid Waste and Solid and Hazardous Waste Exclusions. https://www.epa.gov/hw/criteria-definitionsolid-waste-and-solid-and-hazardous-waste-exclusions.
Van Ewijk, S., Stegemann, J.A., 2016. Limitations of the waste hierarchy for achieving absolute reductions in material throughput. J. Clean. Prod. 132, 122–128.
Veleva, V., Bodkin, G., Todorova, S., 2017. The need for better measurement and employee engagement to advance a circular economy: lessons from Biogen's "zero waste" journey. J. Clean. Prod. 154, 517–529.
Williams, I., 2015. Hierarchy history. CIWM J. 2015 (February), 20–22.
Women in Informal Employment: Globalization and Organizing (WIEGO), 2018. National Solid Waste Policy. http://www.wiego.org/sites/default/files/resources/files/Pereira-Brazilian-Waste-Policy.pdf.
World Economic Forum, 2016. Circular Economy. https://www.weforum.org/globalchallenges/projects/circular-economy/.
Young, C.Y., Ni, S.P., Fan, K.S., 2010. Working towards a zero waste environment in Taiwan. Waste Manag. Res. 28 (3), 236–244.
Zaman, A.U., 2014. Measuring waste management performance using the 'Zero Waste Index': the case of Adelaide, Australia. J. Clean. Prod. 66, 407–419.
Zaman, A.U., 2015. A comprehensive review of the development of zero waste management: lessons learned and guidelines. J. Clean. Prod. 91, 12–25.
Zaman, A.U., Lehmann, S., 2011. Challenges and opportunities in transforming a city into a "zero waste city". Challenges 2 (4), 73–93.
Zaman, A.U., Lehmann, S., 2013a. The zero waste index: a performance measurement tool for waste management systems in a 'zero waste city'. J. Clean. Prod. 50, 123–132.
Zaman, A.U., Lehmann, S., 2013b. The zero waste index: a performance measurement tool for waste management systems in a 'zero waste city'. J. Clean. Prod. 50, 123–132.
Ziout, A., Azab, A., Atwan, M., 2014. A holistic approach for decision on selection of end-of-life products recovery options. J. Clean. Prod. 65, 497–516.

Further reading

Leigh, N.G., Lee, H., 2019. Sustainable and resilient urban water systems: the role of decentralization and planning. Sustainability 11 (3), 918.

CHAPTER SEVEN

Application of advanced technologies in managing wastes produced by leather industries—An approach toward zero waste technology

Mohammed Junaid Hussain Dowlath[a], Sathish Kumar Karuppannan[a], Pratheeka Rajan[b], S.B. Mohamed Khalith[a], Suriyaprakash Rajadesingu[a], and Kantha Deivi Arunachalam[a]

[a]Center for Environmental Nuclear Research, Directorate of Research, SRM Institute of Science and Technology, Kattankulathur, Tamil Nadu, India

[b]PG and Research Department of Biotechnology, Women's Christian College, Chennai, India

Contents

1 Introduction 144
2 Processing steps involved in leather production 145
  2.1 Pre-tanning process 145
  2.2 Tanning process 146
  2.3 Post-tanning 147
  2.4 Finishing 147
3 Overview of leather industry waste 147
  3.1 Solid wastes 149
  3.2 Wastewater 150
  3.3 Volatile organic compounds 151
4 Zero-waste approach 152
5 Technological solutions for the challenges in leather industry 153
  5.1 Technological innovation for cleaner production 153
  5.2 Solid waste treatment and management 158
  5.3 Liquid waste treatment and water management 165
6 Conclusion 169
Important websites 171
References 171

Concepts of Advanced Zero Waste Tools
https://doi.org/10.1016/B978-0-12-822183-9.00007-6

1 Introduction

Waste management has become a global issue. The population growth, urbanization, industrialization, developing economy, and the rise in community standards greatly influence the waste generation (Minghua et al., 2009; Guerrero et al., 2013). All areas of our society like agriculture, transportation, industry, construction, mining, energy, and consumers produce waste. Industrial wastes are generally produced from diverse industrial operations, and as a result, the quantity and toxicity of waste emitted from industrial activities vary with the industrial operations. Industrial wastes can be classified based on their physical characteristics as solid wastes, wastewater, and air discharges.

The leather industry is one of the oldest industries that deliver a large range of consumer products such as shoes, belts, bags, garments, and so on. It is considered as the second oldest profession in the world (Council for Leather Exports, 2020). The leather industry has been the main player in the global trade market. Statistically, the production of world leather is about 574.2×10^6 kg from bovine animals in the year 2014 (Food and Agriculture Organization of the United Nations, 2016). It has grown into an indisputable industry with great economic importance. China is the leading manufacturer of leather goods followed by various countries such as India, Brazil, Vietnam, Italy, and so on. In India, the leather industry has a prominent role in the Indian economy with an export value of US$ 5.69 billion during 2018–19 with about 89% of production from just three states, namely Tamil Nadu, West Bengal, and Uttar Pradesh (Council for Leather Exports, 2020). However, we cannot deny the fact that the economic growth and development of any nation will result in many adverse effects on the environment such as generation of waste, deforestation, emission of greenhouse gases, and so on (Hussain et al., 2020).

The foundation of the leather industry is to convert the waste from agriculture and food industries (meat, dairy, and wool processing) into beneficial, durable, and economically important materials. Various livestock are the sources of leather production with the cattle contributing to about 65%, sheep for 15%, pigs for 11%, and goats for 9% (UNIDO, 2010). The leather industry could have been considered as an initiative for zero-waste technology, which clears out waste from meat processing, wool rearing, and dairy industries by converting the raw skin or hides of the dead animals into fine leather products. But perhaps the waste generated in terms of foul odor, solid waste, and excessively polluting liquid discharges in the process of converting rawhides and skins into leather brings this industry into the list of one of the most polluting industries. The average wastewater release is about 1.5×10^{10} kg per day. Solid waste generation from tannery operations is estimated to be 6×10^{9} kg per year (Rajamani et al., 2009). Scientific vision to address various problems is gaining massive heights as science and engineering move toward a new era (Hussain and Mishra, 2018). The leather industry is developing constantly with progressive changes in its processes, especially in the last two decades.

The new regulations concerning leather production drive leather production more ecological as well. Currently, nearly all the countries including the developing countries have implemented pollution control laws similar to that of the USA, UK, EU, and other developed nations (Hu et al., 2011). After the implementation of new legislations with regard to the environment, with some legislation particularly for the leather industries, new technologies and machinery were introduced for cleaner leather production and as a result, the industry found its successful advancements in the environment means as well as ethical, social, and economic sustainability. Whereas in some countries like the USA, the UK, Germany, the leather processing industries are being shifted to other countries due to the strict laws and environmental guidelines (Genovese et al., 2014; Savino et al., 2015). It is important to notice that the Indian leather products are largely exported to these three countries sharing 15.70, 11.58, and 10.50%, respectively, of India's export (Council for Leather Exports, 2020).

2 Processing steps involved in leather production

The ultimate aim of any leather industry is to transform animal skins/hides into a stable material from which leather products are made for meeting various purposes of people. The process of leather manufacturing has four major sub-processes to obtain the finished leather: pre-tanning process, tanning wet finishing, and finishing. Another additional step, known as surface coating, is done (Tünay et al., 1995; Cooman et al., 2003).

2.1 Pre-tanning process

It is also called as Beam house operation, where it is done to clean and condition the rawhides or skin that cause a large amount of wastewater to be produced. This process involves various stages such as soaking, trimming, deliming, bating, pickling, and degreasing (Tünay et al., 1995; Cooman et al., 2003). During the soaking step, raw skins/hide are treated with water and small quantities of imbibing substances to rehydrate the skin proteins, to solubilize the denatured proteins, to hydrate, and to open the contract fibers of the dried skins, to remove the salt used in the preservation step and to remove the residues of blood, excreta, and dirt attached to the skin (Silva et al., 2012). After skinning at the slaughterhouse, the rawhide contains excessive meat, fleshing, and fats attached to the skin. To remove this, fleshing is done where extra tissues are removed. Fleshing is followed by trimming and unhairing that is done by a liming process in which hair and epidermis are chemically dissolved using an alkaline mixture of lime and sulfide. This step produces wastewater that contains very high COD value. Further to remove the hair remnants and to degrade proteins, the hides/skin are delimed by neutralizing with acid ammonium salts and treated with enzymes, similar to those found in the digestive system. To gain softness and flexibility in the leather, bating is done. Contradictorily, this step loads ammonium salts to the wastewater generated. Pickling is the next step in which the acidity of the hides is increased up to pH3 by adding acid liquor and salts. In this step for preserving the hides/skin, 0.03–2 weight percentage of fungicides and bactericides are added. Degreasing is the final step in the pre-tanning processes and is used to remove the excess of fatty substances using organic solvents/surfactants.

2.2 Tanning process

Tanning is a predominant and most important process for leather making. In this process, collagen fibers are stabilized by action of cross-linking. The tanned hides/skin are intermediate products that are tradable. Tanning agents can be categorized into three main groups, namely, mineral (chrome) tanning agents, vegetable-tanning agents, and alternative tanning agents such as syntans, aldehydes, and oil-tanning agents. Chrome tanning is the most common type of tanning process used in most of the tanning industries in the world. Chrome tanned leather is characterized by top handling quality, high hydrothermal stability, user-specific properties, and versatile applicability. Waste chrome from leather manufacturing, however, poses a significant disposal problem. Vegetable tanning has a limitation as it produces comparatively thick, light brown leather that tends to darken when exposed to natural light. Without specific treatments, the vegetable-tanned leathers will have low hydrothermal stability, inadequate water resistance, and are hydrophilic. Tanning with organic tanning agents can produce mineral-free leather with high hydrothermal stability similar to that of chrome-tanned leather. However, leather that is organically tanned is more filled and hydrophilic than chrome-free leather, with equally high hydrothermal stability (Karabay, 2008).

2.3 Post-tanning

Post-tanning operations involve neutralization and bleaching, followed by retanning, dyeing, and fatliquoring. Specialized treatment may also be performed to increase certain physical/chemical properties such as water repellence or resistance, oleophobicity, gas permeability, flame retardancy, abrasion resistance, and anti-electrostatic properties to the final leather product.

2.4 Finishing

The crust that is obtained after the retanning and drying process has to undergo thorough finishing operations. Such operations intend to make the hide softer and to mask the minute errors. The hide is subjected to organic solvent treatment, a water-based dye, and varnish. Environmental issues are also related to the finishing chemicals, which can also reach the wastewater generated.

3 Overview of leather industry waste

The industrial waste generated from developing countries, such as China and India, seems to be rapidly increasing, whereas in developed countries, the industrial waste is in a stable condition and forecasted to be decreasing in the future. The underlying problem is that collection, segregation, and recycling of waste in developing countries are very low when compared to developed countries. This shows that developing nations are facing more challenges related to industrial waste.

Leather-processing technology has evolved naturally as a traditional practice to an industrial process. The leather manufacturing processes produce considerable quantum of solid, liquid, and gaseous wastes (Fig. 1). If managed correctly, waste can be a valuable resource (Ngoc and Schnitzer, 2009). For processing, one metric ton of rawhide, 200 kg of tanned leather, 200–250 kg of tanned leather waste, 190–350 kg of untanned waste, and 50,000 kg of wastewater is produced (Fig. 2), which means just 20% of the raw material becomes finished leather (De Aquim et al., 2010).

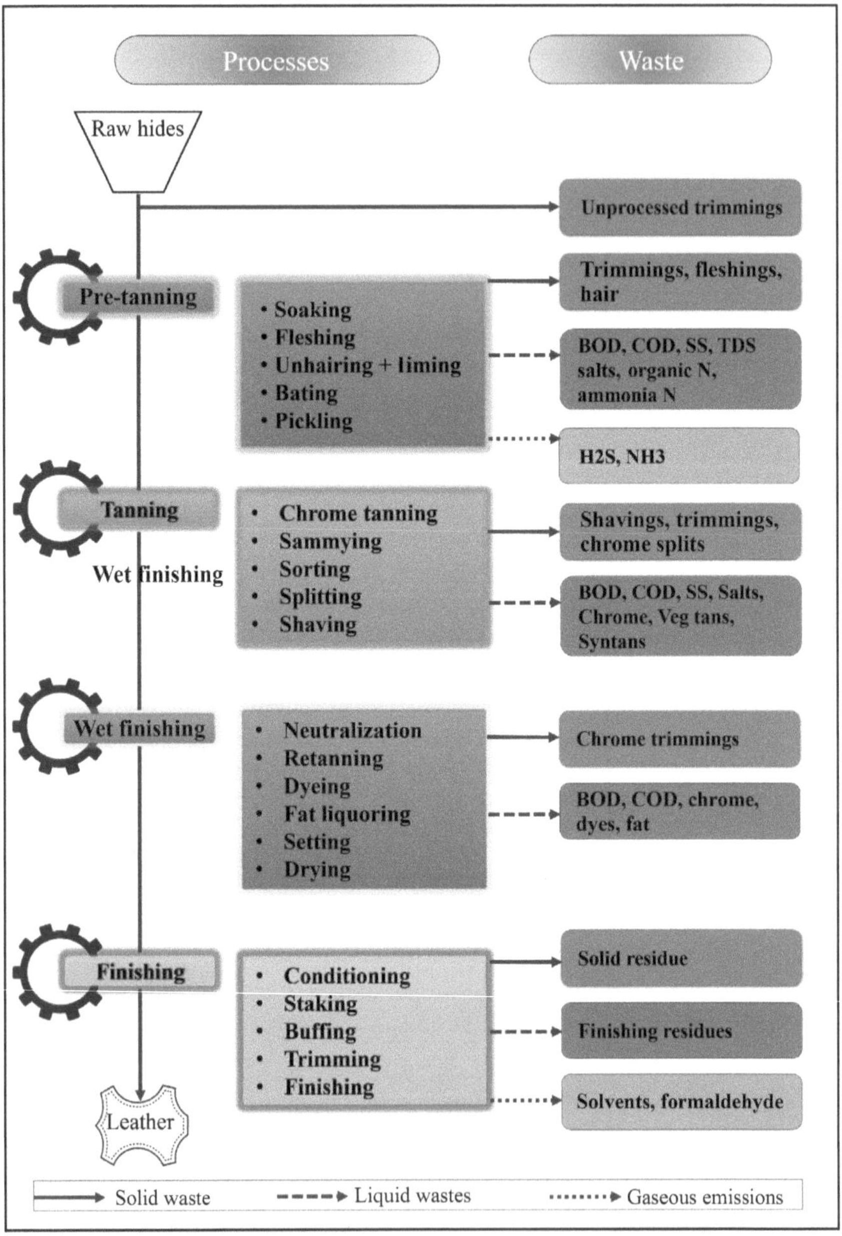

Fig. 1 Waste generated during leather processing (UNIDO, 2011). *(Modified from UNIDO, 2011. Introduction to Treatment of Tannery Effluents. [online]. United Nations Industrial Development Organization, Vienna. Available at: https://www.unido.org/resources/publications/creating-shared-prosperity/agribusiness-and-rural-entrepreneurship/introduction-treatment-tannery-effluents [Accessed 25 April 2020].).*

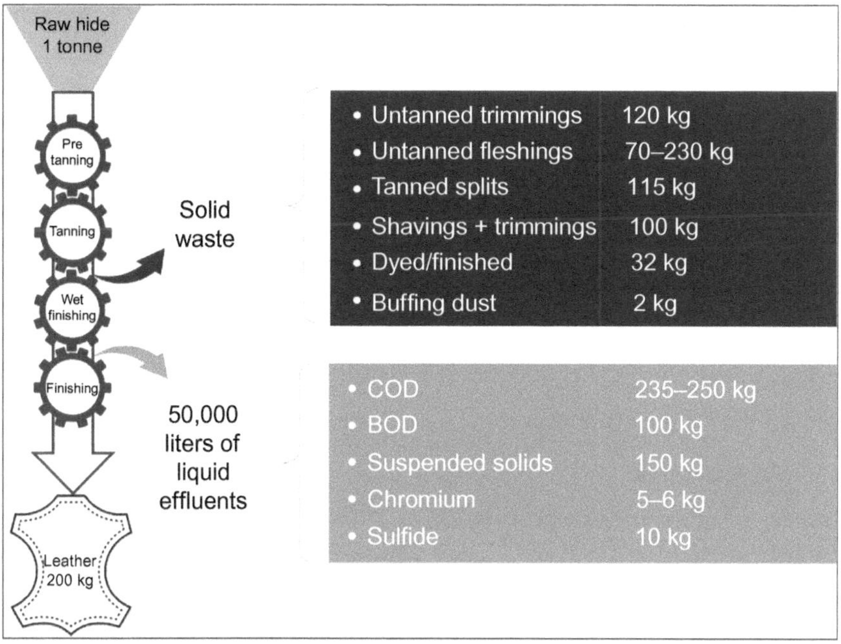

Fig. 2 Environmental impact of leather processing (Chakraborty, 2003). *(Modified from Chakraborty, R., 2003. Ecofriendly solid waste management in Leather Industry. Environ. Biotechnol. Newsl., Univ. Kalyani, 3 (2).)*

3.1 Solid wastes

The intrinsic properties of the steps involved in leather processing and properties of the chemicals used are also the reasons for a certain amount of solid wastes being produced. These wastes are a threat to the ecosystem in the vicinity of tanneries (Mwinyihija, 2010).

Solid wastes from the leather industries are unavoidable. Solid wastes generated during the production of leather can be classified as follows:

i. Wastes from untanned skin/hides (e.g. trimmings and fleshing wastes)
ii. Wastes from tanned leather (e.g. shaving wastes and buffing dust)
iii. Wastes from dyed and finished leather (e.g. trimmings)

Research studies reveal that 80% of solid wastes are generated during pretanning operation, while 20% of the wastes are produced during posttanning processes (Ozgunay et al., 2007). Apart from the fact that leather degrades slowly, treating it with various chemicals in the process of tanning makes it more rigid and resistant toward chemical, thermal, and microbial degradation (Hagerman, 1980; Han et al., 2003). Because of the nonbiodegradability of tanned leather, solid waste management system finds difficulties in dealing with the sludge generated from tanneries (Dhayalan et al., 2007; Lofrano et al., 2008), and this directly causes a negative impact on agro-based activities and degrades groundwater system (Mwinyihija, 2012). The addition of pesticides for skin/hides conservation during transport also adds to the problem. Bovine and ovine hair is obtained as a by-product from the tanneries during hair-recovering unhairing process, and it is assessed that about 5% of dry hair is recovered based on the rawhide weight. Mostly, the waste generated by the leather industries is disposed of through landfill or incineration processes. These methods of disposal possess economic and environmental losses due to the steps involved in this such as transportation. Hence, the effective reutilization is seriously required.

3.2 Wastewater

During the entire tanning process, a huge amount of water and pollutants are discharged (Kaul et al., 2001). Large volumes of wastewater contaminated with different chemicals and organic matters pose more challenges in reality in comparison with treating the wastewater. To process one metric ton of wet salted hides, 40 m^3 of water is required, and out of 452 kg of process chemicals used, only 72 kg of leather is taken and the remaining 380 kg is discharged (Catalina et al., 2007). The details of water consumption for several steps and the characteristic pollutant loads for each operation are presented in Table 1.

Pre-tanning and tanning operations contribute about 90% of the total pollution caused by the leather industry. The pre-tanning operation causes differences in pH levels, thereby increasing the chemical oxygen demand (COD), total dissolved solids (TDS), chlorides, and sulfates in wastewaters discharged by tanneries. The lime and sulfide utilized during dehairing process contribute 84% of biochemical oxygen demand (BOD), 75% of COD, and 92% of suspended solids (SS) from a leather manufacturing industry. The utilization of sulfides not only increases negative consequences on the environment but also affects the efficacy of wastewater treatment plants. During chrome tanning, the poor chromium uptake causes material wastage and results in ecological imbalances. The post-tanning process also leads to significant modifications of COD, TDS, and heavy metal pollution (Thanikaivelan et al., 2000, 2005; Dixit et al., 2015). It has been assessed that more than 55,000 ha of land has been polluted by tanneries, and about 5 million people are affected directly or indirectly by the low-quality drinking water and social environment (CSIRO, 2001).

Table 1 Quantity of wastewater generated and pollution load of each step during processing 1 ton of skin/hide.

	Processing operations load kg/t hide					
Pollution load	**Soaking**	**Unhairing/ liming**	**Deliming and bating**	**Chrome tanning**	**Post-tanning**	**Finishing**
Wastewater generated (m^3 or kiloliter)	9.0–12.0	4.0–6.0	1.5–2.0	1.0–2.0	1.0–1.5	1.0–2.0
Suspended solids	11–17	53–97	8–12	5–10	6–11	0–2
COD	22–33	79–122	13–20	7–11	24–40	0–5
BOD	7–11	28–45	59	2–4	8–15	0–2
Chromium	–	–	–	2–5	1–2	–
Sulfides	–	3.9–8.7	0.1–0.3	–	–	–
NH_3–N	0.1–0.2	0.4–0.5	2.6–3.9	0.6–0.9	0.3–0.5	–
Total Kjeldahl nitrogen	1–2	6–8	3–5	0.6–0.9	1–2	–
Chlorides	85–113	5–15	2–4	40–60	5–10	–
Sulfates	1–2	1–2	10–26	30–55	10–25	–

Reprinted from Dixit, S., Yadav, A., Dwivedi, P.D., Das, M., 2014. Toxic hazards of leather industry and technologies to combat threat: a review. J. Clean. Prod. 87, 40, Copyright (2014), with permission from Elsevier.

Attempts have been made to combat the liquid discharges of the tanneries to near-zero levels because leather processing is primarily associated with purification of a multicomponent, skin, to obtain a single protein, collagen (Dixit et al., 2015).

3.3 Volatile organic compounds

Tanneries release various gaseous pollutants such as ammonia, amines, aldehydes, hydrogen sulfide and volatile hydrocarbons into the atmosphere as effluents (Famielec and Wieczorek-Ciurowa, 2011). During deliming, unhairing, or drying processes, ammonia is emitted, and during liming/unhairing and subsequent processes, emissions of sulfides occur. Particulate emissions contain chromium, which may occur because of the reduction of chromate or from handling basic chromic sulfate powder or even from the buffing process (EPA, 1982; El Samra et al., 2007). Thus, significant quantities of volatile organic compounds (VOC) are released during various processes of tanneries which pose a threat to the atmosphere if not properly controlled.

These manufacturing practices are responsible for polluting the environment and reputation of the leather industry.

4 Zero-waste approach

For addressing the aforementioned problems, solutions based on zero-waste are being approached. Zero-waste is a philosophy of the 21st century to reduce the generation of waste that has no benefits to human society and the environment. The term zero-waste was first used by Dr. Paul Palmer in 1973 for recovering resources from chemicals (Song et al., 2015). By adopting various new technologies, through innovations and inventions, the waste is being either reduced or completely nullified by converting the waste into some or other value-added products. The zero-waste approach is a step toward the UN's Sustainable Development Goals 2030. This approach will emphasize the goals 03, 06, 07, 08, 09, 12, 13, 14, and 15. Sustainable development is a mix of social, economic, and environmental development (Ueda et al., 2009). Many researchers believe that sustainable development for the leather industry is impossible unless it grips groundbreaking tactics to prevent and alleviate the challenges of pollution (Sarkis et al., 2010; Colicchia et al., 2011; Shamma and Hassan, 2013; Deng, 2015; Hosseinpour et al., 2015).

Technologies supporting the zero-waste approach are very well established in leather industries and are found to be effective in reducing the pollutions and hazards caused by waste. This chapter further discusses what are the measures followed by the leather industries with the zero-waste approach. This includes alternative methodologies in leather processing to reduce waste production such as changing the conventional instruments to modern machinery, the efficient use of resources in all possible directions, conversion of generated waste into economically important or pollution less products, remediation of waste, and effective policies and strict regulations (Global Alliance for Incinerator Alternatives, 2020). These approaches signify the 3 R's – Reduce, Reuse and Recycle – and also, the new 3 R's – Refuse, Rethink (Reinvent) and Replace – which promotes ecological balance. This approach in India is addressed by the R&D efforts made by CSIR- Central Leather Research Institute located in Chennai with its allies. At present, the environment is the main field of research carried out by the leather research institutes and universities. The leather dealt with cleaner production and waste management is a major issue for sustainable development of the leather industry (Rajamani et al., 2009).

The processes involve a variety of novel scientific approaches such as enzyme assisted pre-tanning and tanning processes, increasing the chromium exhaust by introducing new tanning auxiliaries, enhanced waste treatment systems, and management of waste by utilizing it in various other applications such as for production of feed, fuel, and so on.

5 Technological solutions for the challenges in leather industry

In the last two decades, the leather industry is facing various issues related to environmental concerns. The most significant approach in preventing environmental pollution is the idea that prevention is better than reuse, reuse is better than recycling, and recycling is better than disposing the wastes (Rao et al., 2003). This situation has led the leather industries to find solutions based on greener technologies (Gupta and Babu, 2009; Sundarapandiyan et al., 2010, 2011; Krishnamoorthy et al., 2012), in other words, cleaner production. Dixit et al. (2015) suggests two significant methods to reduce the impact on the environment caused by the leather industries. The two methods are as follows:

(a) Cleaner technologies or low waste: this method suggests innovation and improvement for cleaner production and minimizes the use of hazardous chemicals.

(b) Treatment of wastewater: In this, the wastewater is treated in an eco-friendly processing method.

In this chapter, these methods are seconded with slight modifications as

(a) Technological innovation for cleaner production where the production in the leather industry has less harm to the environment

(b) Treatment and management of waste in which waste generated by the leather industry is treated and reutilized as a raw material for some other processes or value-added products.

By following these methods, the negative impact posed on the leather industries can be reduced.

5.1 Technological innovation for cleaner production

In recent times, the prevention or minimization of leather industry pollutants at the source by following in-process controls is gaining attention (Rao et al., 2003). Factors affecting the tanners to adopt cleaner production technologies are as follows:

(a) Recognized minimization of effluents in terms of both quality and quantity
(b) Proven economic advantages such as material saving and cost reduction
(c) Minimum investments on machinery and effortlessness during the application
(d) Trade benefits on account of enhanced environmental placing in the world market.

The growing pressure on the leather industries to adopt cleaner production technologies normally originates from environmental essentials like the necessity to follow defined discharge norms, reduction of treatment costs, or follow the occupational safety and health (OSH) standards. In industrialized countries, such technologies that meet the requirements are adopted on a large scale. On the wider scope, solutions based on biotechnology have been followed in the leather industries for many years, from the beginning of enzyme usage. Enzymes can be employed in all the steps of leather production, with some exceptions during the tanning process. Currently, methods of biological origin are being followed with comparative achievement in all the functions of a tannery.

5.1.1 Pre-tanning process

Pre-tanning processes, such as soaking, liming, and dehairing, are the steps in which huge amounts of solid waste are generated. Fifty percent of the solid waste obtained during these stages comes from fleshing processes, 20% originate from trimming, and 25% originate from hair and wool removal.

Enzymatic soaking process

Enzymes used in soaking process quicken the process of eliminating hyaluronic acid and enhances the quality by effectively rehydrating the skin, degrading the fat, and improving the elimination of carbohydrates and fatty components (Cavello and Cavalitto, 2013; Dettmer et al., 2012; Nazer and Siebel, 2006). The fats and proteins of non-collagen origin which are solidified interrupt the fibers and are attacked, thereby shortening the production period. A soaking process using TanG, LimeG and FP50 enzymes was recommended by Ma et al. (2014). In another study, researchers suggested that the use of ultrasound for the soaking of skin would reduce the water usage and time consumed (Morera et al., 2013). By using enzymes, BOD/COD ratio was brought to 0.89 that indicates the biodegradable nature of the waste. Jayanthi et al. (2019) reported 30% minimization of soaking duration while using carbohydrase, an enzyme isolated from *Aspergillus terreus* 5409 for pre-tanning operations like soaking unhairing and fiber opening.

Enzymatic liming process

By using enzyme-assisted processes, the high proportions of lime and sulfide generated by liming process can be reduced. Proteolytic and catalyzing enzymes laterally with minor quantities of lime and sulfide are used in dehairing which generally breakdown the proteins. This enzyme combination is coated onto the flesh side, which causes the loosening of hair by selectively breaking down the cementing substances in the hair follicle, resulting in intact hairs (Thanikaivelan et al., 2004).

Enzymatic degreasing process

The use of enzymatic products is essential during the degreasing stage of the tanning process. An enzymatic preparation needs to have properties such as proteolysis, lipolysis, and emulsification to be an efficient degreasing agent. The use of alkaline lipase along with pancreatin and proteinase increased the degreasing by softening the pigskin (Zhang and Zhang, 1982). During soaking and bating, it is proved that the mixture of enzymes with suitable surfactants has a synergistic effect that enables the optimal degreasing of rawhides, wet blues, and pelts (Pfleiderer, 1983). Lipase enzymes of bacterial or fungal origin can be used to degrease the animal skin/hides under acidic or alkaline conditions. Enzyme-assisted degreasing technology is not only efficient compared to conventional processing but also is a time-saving one. Enzymatic degreasing is considered to be an effective subordinate for fighting against the pollution issues caused by the use of detergents and solvents (Kamini et al., 1999). A recent study attempted and accomplished a 100% surfactant-free beamhouse operation by using lipase enzymes for degreasing the pigskins (Li et al., 2020).

Enzymatic dehairing of skin/hide process

Senthilvelan et al. (2012) developed an enzymatic unhairing process by using protease enzyme instead of sulfide. This enzyme has its origin from a *Bacillus* sp. The enzyme activity was optimized to pH 11.0, at 45C for 16 h. This finding showed the achievement of complete dehairing by using 2% alkaline protease for 16 h. This method substantially reduced the contamination levels to BOD – 62.8%, COD – 79.0%, TDS – 82.5%, and TSS – 88.2% when compared with the control. Dayanandan et al. (2003), isolated enzymes from *Aspergillus tamarii* and used them for dehairing goat skins.

The obtained leather was of good quality with comparative strength properties and also reduced the pollution levels of BOD, COD, TSS, and TDS to 50%, 40%, 20%, and 60%, respectively. Galactosidase enzyme isolated from *Aspergillus terreus* was used for opening of fibers in the dehaired goat skins. Fiber opening was achieved at 0.5% concentration of enzyme treatment for 30 min. Owing to enhanced fiber opening, this treatment increased the chromium uptake of the leather, thereby reducing the pollution and producing the leather with better physical properties (Durga et al., 2016).

In another attempt for loosening the hair, commercially available alkaline protease enzymes from bacterial origin were used. The complete hair removal was optimal at pH 8.0 with no addition of lime for dehairing (whereas usual pH used to be 12.0 when lime is used). This benefits the environment by reducing the COD by 45% and TSS by 13% (Thanikaivelan et al., 2003). To cut the cost spent on enzymes, Saravanan et al. (2014) proposed the use of sans lime and sodium sulfide and also recommended ideas for reducing the quantity of enzymes to be used. Complete dehairing in goatskins was achieved by using 1% enzyme concentration with no addition of sodium sulfide (Saravanabhavan et al., 2003a). To achieve complete removal of hair from sheepskins, a mixture containing 1% enzyme concentration of 1% and sodium sulfide at 0.25% was used (Aravindhan et al., 2004). Aravindhan et al. (2007) suggested that the use of enzymatic dehairing and pickle less chrome tanning can be eco-friendly options by reducing the COD to 67% and also cost-effective.

5.1.2 Tanning process

In traditional chrome tanning procedures, material utilization was only 40–70% and the dumping of left-over wastes is a matter of concern for environments due to the pollutions caused by the wastewater discharges and management of solid waste with chrome residues (Bacardit et al., 2014). Essentially improved chrome tanning salts, usage of chromium syntan, tanning process with high exhaustion, recycling/reusing techniques would consistently help to increase the use of chrome resulting in the decrease of chrome release (Suresh et al., 2001; Thanikaivelan et al., 2005; Gupta and Babu, 2009; Morera et al., 2006; Sundarapandiyan et al., 2011; Frendrup, 1995; International Union of Environment Commission, 2018; Rao et al., 2002, 2002a).

Enzyme-assisted three-step tanning

An enzyme-driven three-step tanning process applicable to hides and skins has been developed that involves enzymatic unhairing, enzyme-based fiber opening, and pickle-free chrome tanning at pH 4.0–8.0. In addition, an approach that uses plant-based polyphenol compounds for tanning instead of chromium at pH 8.0 has also been suggested for the three-step tanning process. The three-step tanning technique reduces chemical consumption by 98% and discharge of chemicals by 88%. Moreover, this process also provides economic benefits in terms of water reduction, reduced power consumption, and increased area yield (Aravindhan et al., 2004; Saravanabhavan et al., 2003b; Saravanabhavan et al., 2004; Thanikaivelan et al., 2003).

High exhaust chrome tanning

Biotanin is a protein-based tanning medium prepared using the fleshing wastes generated by tanneries. In the tanning process, chrome tannin exhaustion was improved from 67% to 92% by applying 2% of Biotanin on skins/hides. This also reduced the COD levels by 58%. The use of basified salts instead of conventional chromium salts during tanning which helps in the recycling of basified tanning float liable on the quality of grain of leather. Water saving of up to 90% was achieved by this method (Morera et al., 2011). In commercial leather industries, a novel method of high exhaust pickle-tan chrome tanning has been successfully developed and proved. The combination of basification with magnesium oxide, modification by tartaric acid at 3% concentration resulted in 93.51% exhaustion at 35°C (Nashy and Eid, 2019). Normally pickling will be done using 8–10% sodium chloride with pH maintained between 2.5 and 3.0 and basic chromium salt at a concentration of 8%. But in this method, normal pickling using anhydrous sodium sulfate at 5% concentration is employed followed by chrome tanning with combined effects of aluminum syntan and 5% of basic chromium sulfate (BCS) at a pH of 3.0–3.2 (Rao et al., 2003). By this method, during tanning, about 90% in chrome absorption levels are attained which reduces the levels of BOD, COD, and TDS loads in the wastewater. Owing to the enhanced uptake of chromium and the presence of aluminum, this method of tanning additionally reduces the amount of chemicals used during post-tanning operations. It reduces about 30–40% of retanning agents, 5–7% of fatliquors, and 10–20% of dyes.

Another high chrome exhaustion tanning method by using novel nanoparticle dispersion (NPD) was suggested by Kanagaraj et al. (2015). The application of 4% NPD during chrome tanning showed the improvement in exhaustion up to 94%. The obtained leather was found to be of better quality with comparative strength properties and also proved to be an eco-friendly method for leather manufacturing (Kanagaraj et al., 2015). To improve the exhaustion of chrome tanning, another waste released from the slaughterhouse known as fibrin hydrolysate was used in aiding chrome exhaust. Fibrin hydrolysate of 3% was found effective in increasing the exhaustion of chrome tannins from 70% to 91%. Owing to the increased uptake of chrome during the tanning process, pollution levels like BOD, COD, TDS and TSS dropped to 50.80, 54.30, 17.80, and 17.90%, respectively. Researchers suggest that this cleaner technology can reduce about 18% of the solid waste generated by the leather industries.

5.1.3 Post-tanning process

Dyeing process by using polymer

The chemical dyes that do not exhaust completely during the dyeing process and get released as effluent from tanneries is a cause of environmental concern. A suitable technology is needed in solving the issue of degrading the dyes completely, in the leather industry (Asad et al., 2007; He et al., 2004; Elisangela et al., 2009). To improve the exhaustion of dyes, Nanoparticle polymer (NPP) is used. During the dyeing process, by applying 2% of NPP on the leather improves the dye uptake up to 99.10%. The fact that NPP enhanced the uptake of dye was proved by studies based on the kinetic model and color measurement and FTIR data supporting the result by showing the bonding of hydrogen in between the various elements of secondary structure. The application of graft copolymer with 10% concentration resulted in enhanced dyeing ability and increased dye exhaustion up to 96% at an optimized condition of pH 5.0 and 60°C for 80 min. This serves as a cleaner dyeing process for producing eco-friendly leather (Kanagaraj et al., 2016).

5.2 Solid waste treatment and management

The majority of the developing and underdeveloped countries dump their solid waste without following safe practices like landfilling (Dowlath et al., 2020; Hussain, 2019). Solid wastes possess a key problem for the leather industries in terms of both their diversity and size. Solid waste generally constitutes wastes like blood, natural fats, proteins, salts, and other wastes. These wastes possibly can be recycled and reutilized and can serve as raw material for various other useful products (International Union of Environment Commission, 2018; Colak et al., 2005). Various factors like species of the animal, the conditions in which it is bred, practices followed in slaughterhouse, preservation conditions, stages in leather processing and the usage of chemicals affects the diversity and amount of solid waste generated. In the process of leather manufacturing, global chrome shavings waste generated per year is estimated to be around 0.8 million tons constituting about 75% of solid wastes. Effective utilization for these resources is important in terms of both environmental concern and to prevent the loss of resources.

5.2.1 Treatment of wastes

The chrome shavings and trimmings produced after the tanning process are the major solid wastes that are of great concern. The leaching of Chromium into the groundwater, conversion of Cr^{3+} into Cr^{6+}, emission of radicals, the formation of ammonia and hydrogen cyanide are some of the problems faced by the leather industries on following conventional wastewater disposing methods (Rai et al., 1989). Storing this kind of wastes in some places is also a potential risk and very costly. Various studies are being conducted in search of safe methods of disposal, recovery of chromium and proteins released in wastes and for their purpose in different industrial applications. To fulfill the increasing pressure, opting for a cleaner or low waste technologies, processing of the waste generated by the tanneries into valuable resources will be economically benefitting the leather industry.

Incineration

Even though the chrome shavings discarded from the tanneries are incinerated or buried underground, it can still cause environmental degradation by a large amount of chromium present in it. Shavings contain a large amount of proteins ranging between 78.6 and 75.2% and chromium oxide in the range of 4.4–4.3%. By using wet air oxidation studies have been carried out to degrade the organic material present in the shavings and to isolate the chromium alone from the oxidized liquor. Various researches were made to reduce the release of toxic gases during incineration of chromium-containing wastes and to effectively remove it using the scrubber. The results showed that the concentration of different gas was within the permissible limits and ash obtained in this process can be used for preparing bichromate (Pati et al., 2014). To reduce the burden of chromium on the ecosystem, Sekaran et al. (2007) suggested to incinerate the tannery waste by air starved thermal incinerators and to solidify the calcined waste which results is 99.15 of metal ion fixation. The incineration of organic chrome shavings without the oxidation of Cr^{3+} into Cr^{6+} by starved air thermal incineration helped in improved retrieval of energy (Swarnalatha et al., 2009).

Alkaline hydrolysis

Solid wastes after the tanning process have an organized structure appearing as fibers, closely held with each other. Alkaline hydrolysis is a method alternate for detanning the chrome wastes. Potential extraction of important products like collagen hydrolysate and chrome cakes at various processes was made possible by a three-step hydrolysis process. These products had helpful utilization in the leather industry (Pati et al., 2014; Mu et al., 2003; Saha et al., 2003; Ferreira and Almeida, 2011). In another study, the presence of peroxochromates production was found in situ while oxidizing at the alkaline condition. These peroxochromates are found to create fractional hydrolysis to the fibers of the collagen which accelerates the extraction of gelatin (Cot et al., 1999). At the same time, alkaline hydrolysis in the presence of sodium hydroxide allowed the retrieval of metal salts at the solid state and the proteins during the liquid phase. The proteins obtained in the liquid phase are lyophilized to recover solid proteins that are rich in nitrogen concentration and least in chrome content. To remove the chromium present in the leather waste, the use of organic chelates as cleansing agents and potassium tartrate as an extracting agent was found to be very effective in the alkaline medium. The use of hydrogen peroxide improved with ultrasound in oxidative dechroming was found to be efficient in removing chrome up to 99%. This will be significantly useful for the use of gained collagen products (Malek et al., 2009). In another attempt for treating the tannery wastes, fleshings subjected to alkaline hydrolysis yielded collagen hydrolysate which also has notable applications (Castiello et al., 2006). By alkali hydrolysis attempts are also made in the process of converting the keratin-rich wastes like hair, horns, hoofs, poultry feathers, and so on into keratin-rich hydrolysate.

Biological degradation

Enzymatic processing of the chromium-containing tanning waste is a feasible method and avails a hydrolysate yield of 50–60%. Solubilizing the protein is effective by using proteolytic enzymes which are active at moderate temperatures and due to alkaline pH during the reaction, the chromium stays insoluble. The chemical elements present in the isolated products depend on the factors like treatment method, the actual composition of chrome containing waste. A two-step treatment in which both the use of alkali and enzymes is found to be effective in treating chrome shavings. In the first step, gelatin is isolated using alkali and in the next step, protein is recovered using an enzyme that helps in treating and recycling the chrome cake (Sundar et al., 2011a). Using alkaline protease, various protein-based products with diverse properties can be obtained (Cabeza et al., 1998). The potential utilization of chrome cakes and aerobic biodegradation in an aqueous medium was identified (Dvorackova et al., 2007; Hrncirik et al., 2005; Kupec et al., 2002). Potential organic degradation of chrome shavings by microbial enzymes of *Pseudomonas aeruginosa* origin was found to reduce the hydrolysis time (Sivaparvathi et al., 1986; Ferreira et al., 2014).

Tannins used for preventing the disintegration collagen fibers are discharged as wastes. Organic degradation of these tannins and dyes was done using *Penicillium* sp. and *Aspergillus niger* cultures for about 11 days and successive ozonation for 4 h to improve the degradation rate. Ozonizer at 8 g/30 min concentration with a flow rate of 2 lpm was used for ozonation. The study results showed the degradation efficacy of vegetable tannin, chrome tannins, and dyes to be about 92–95%, 85–87%, and 94–95%, respectively. By using the combined effect of degradation and ozonation chrome tannins can be converted into chromic oxide. This combined effect method is found to decrease the levels of COD, TDS, and TSS (Kanagaraj and Mandal, 2012).

5.2.2 Management of Recovered Material

Animal feed/chicken feed/fertilizer

A step toward zero waste is the wise usage of the tannery wastes in the applications such as feed, fertilizer, or cosmetic additives (Castilhos et al., 2002; Konrad and Castilhos, 2002; Lima et al., 2010). The protein present in the tannery waste is a rich source of protein for poultry feed application (Paul et al., 2013). Hydrolysis of the tannery waste which is tanned using conventional chrome tanning methods can produce gelatin of food-grade quality. Hydrolyzed protein can be utilized as amino acid supplements in animal feed because of the presence of high amounts of nitrogen content (Brown et al., 1996). Wet blue leather waste after chromium extraction through its use in ruminant feed characterizing its nutritional potential for these species. The high crude protein content of the leather residues that undergo chromium extraction indicates that they can be used as a protein supplement in animal feed. The production of leather meal with almost no chromium content (<0.1 ppm) is possible. The technique of chromium extraction of wet blue leather residues, with consequent attainment of leather waste without chrome, makes this material highly degradable in the rumen and highly digestible in the abomasums, providing it with great nutritional potential. The amino acid composition is found to be of higher value when compared to that of feather meal and equivalent to meat meal and in regard to economic affordability is the same as soybean meal. The protein recovered is utilized as a fish feed supplement for *Oreochromis niloticus* (El Boushy et al., 1991; Katsifas et al., 2004; Montoneri et al., 1994).

Fleshings hydrolyzed by pancreatic enzymes could be used as a feed formulation by mixing with other feed ingredients. Delimed fleshings, mixed with *Pediococcus acidolactici* culture, sugar, and common salt, on fermentation at 30 ± 2°C yield silage (protein 7%, fat 8%, and ash 3.6%), which can be used in animal feeds. This can also be used in aquaculture and animal feeds to supplement arginine, phenylalanine, and tyrosine (Bhaskar et al., 2007). Hide fleshings were mixed with carbohydrate residues, and then fermented and inoculated with fungi. The resulting substance was claimed to be a useful added nutrient to other feeds for raising cattle and poultry. The hide powder prepared from limed split waste is used as an animal protein source for poultry feed. For this purpose, the hide powder is preferred to be prepared from limed split waste by reliming at 18–20°C for 20 days, deliming, enzymatic treating, and removing fat by soaking in acetone and dehydrating. The action of fertilizer from fleshings is similar to that of organic manure. It is a slow nitrogen release agent since the final protein breakdown is accomplished with the aid of soil microorganisms. For this reason, it is added to the soil before planting. An enormous quantity of keratins in the form of hairs, feathers, horns, and hoofs are wasted each year. Feather (90% keratin) is used as animal feedstuff in the form of a feather meal. Keratin has been investigated as a protein source for balanced chicken feed. The hydrolysis and solubilization of hair require the breakdown of the structure of the protein keratin, the major hair component (Sundar et al., 2011b).

In leather processing

A new method was evolved for utilizing leather waste in preparing newer types of surfactants from protein hydrolysates. These low-cost surfactants are used in soaking and fatliquoring of leather processing. The application of keratin hydrolysate produced from hair waste improved the exhaustion of chromium in chrome tanning as well as in rechroming processes. During leather processing, keratin hydrolysate can also be used for filling and retanning processes (Karthikeyan et al., 2007). The casein formulation using wet blue shavings is made possible by completely digesting it using alkali protease (Crispim and Mota, 2003). Collagen hydrolysate produced from tannery fleshing has its application in vegetable retannage dyeing and vegetable-chrome retannage dyeing to produce upper leather replacing fillers and syntan. Studies showed a high level of exhaustion and leathers with comparatively better properties (Castiello et al., 2006). The use of collagen hydrolysates during chrome tanning is proved to contribute to increasing the chromium oxide levels of the leather at the end of tanning operations. During post-tanning operations, the synergetic effect of collagen hydrolysate and acrylic-retanning agents exhibited striking properties. During post-tanning operations, due to the synergetic effect of collagen hydrolysate and acrylic-retanning agents, improving the properties of leather grain and enhanced tensile strength by the formation of polyelectrolytes exhibited striking properties. Collagen hydrolysates can also be used for aiding dye exhaustion (Afşar et al., 2010; Aslan et al., 2006; Munoz et al., 2002). Finishing agents for leather can be produced potentially by chemically modifying the gelatin. For preparing chrome tanning agents, Cr^{6+} is reduced by using the chrome shavings which are known to have greater protein contents. And because of the intermediate oligopeptide formations, the resultant product showed high masking (Rao et al., 2002b).

For energy

It is said that energy from the leather industry is sufficient to satisfy the heat energy need of a tannery for processing the leather, known that the energy value of leather wastes is to be about half the value of coal. Therefore, in an attempt to eliminate the solid tannery waste, drying, and gasification, a flammable gas was combined as a source of renewable energy that the leather industry would reuse. Biogas is being produced from leather industry wastes like splits, trimmings, fleshing scrap, and sludges from effluent. Possible production of cost-effective methane has been proved technically. Biogas with enriched nutrients with feasible utilization in agricultural activities can be produced by anaerobically digesting the tannery solid wastes (Gnanamani and Bai, 1992; Dhayalan et al., 2007). Various studies showing the possible utilization of fleshing for producing biodiesel, fatliquor, soap, and so on have been explored (Colak et al., 2005; Ravindranath and Gopalakrishnan, 2010). An attempt was made to produce biodiesel (FAME – fatty acid methyl ester), an alternate for fossil fuels using the fat residues present in the animal fleshing by transesterification. Boiling the fleshing waste with water with rapid grinding and drying released the fat which was then used for producing FAME (Gaidau et al., 2007; Gaidău et al., 2009). Shin et al. (2008) successfully demonstrated the process of producing biodiesel from tannery fleshing by immobilized lipase-catalyst. Fleshing scraps from tanneries on boiling at high temperatures and successive treating with sulfuric acid obtained fats with FAME. In another attempt fats from tannery pre-fleshings were used for biodiesel production. Hence, waste that is contaminating is successfully converted into a biofuel that is eco-friendly and has economic benefits (Colak et al., 2005; Kolomaznik et al., 2009).

Adhesives/cosmetics/films

Chrome tanned leather splits, trimmings and buffing dust can be used to produce collagen, chrome cakes, gelatin, glue and adhesives (Brown et al., 1996; Cantera, 2003; Cantera et al., 2002; Taylor et al., 2003). The methods for the production of glue, sausage skins from fleshings, and other such waste are patented. Animal fleshing serve as raw materials for the manufacture of adhesives in the glue and gelatin industries. The fleshings that are not used to produce glue and gelatin are proposed to have its application such as artificial leather, amino acid production, plastics, and poultry feed. Protein hydrolysates with high purity after demineralization can be used in the production of cosmetic products such as hair sprays, moisturizing lotions, and creams and for medical applications such as in dressings, burns, implant coverings, and so on (Cakl et al., 1998). Classically, products of collagen hydrolysates like animal glue and gelatin have found its usage as in paints, adhesives, flocculation, encapsulation, and fireproof agents. Microencapsulation of drugs, perfumes, essential oils, and various other things has been carried out using gelatin. Gelatins are used to prepare films by enzymatic treatments and glycerol is used as a plasticizer that is used for preparing edible films, coatings, sausage casings, and packing material (Cabeza et al., 1998; Taylor et al., 1998, 1997).

Biological uses

The collagen hydrolysate produced from the shavings can be potentially used as a supporting matrix to immobilize the enzyme organophosphorus hydrolyze. The collagen hydrolysate can serve as a rich source of carbon, nitrogen, and amino acid (Aslan et al., 2007). Cr^{3+} present in chrome waste is a cofactor for insulin and is vital in metabolic functions. This can be used to produce brewer's yeast with enriched chromium (Liu and Yang, 2006). Because of the rich protein content in animal fleshing, it is used as a substrate for producing alkaline protease using *P. aeruginosa* (Kumar et al., 2008).

In leather boards

Chrome tanned shavings can be used in the manufacture of leather boards. Leather boards of good quality with flexibility were manufactured using tanned wastes such as chrome shavings, splits and trimmings along with kraft pulp and cotton (Crispim and Mota, 2003; Gish, 2000).

5.3 Liquid waste treatment and water management

5.3.1 *Treatment of wastes*

Today water purification is a global problem, especially in industrialized countries, where wastewater usually contains organic pollutants, such as dyes released from the tanneries, textile industry, paper production, pharmaceutical industries, and so on (Hussain and Kharisov, 2017). Various developments are being made in the treatment of tannery wastewater. Technologies based on lagoons, bio methanation, upflow anaerobic sludge blanket reactor (UASB), contact filter, and so on have been adopted for anaerobic treatment tannery wastewater (Lefebvre et al., 2006; El-Sheikh et al., 2011). Potential nanomaterials such as TiO_2, ZnO, carbon nanomaterial-based membranes, polymers are used for environmental analysis and elimination of various dyes (Hussain and Kecili, 2019; Bensouici et al., 2016; Karupannan et al., 2020). Aerators with the presence or absence of aerobic microbes are used for post-treatment procedures of tannery wastewater (Farabegoli et al., 2004; Ganesh et al., 2006). Conventional treatment procedures are replaced by the usage of membrane technologies like reverse osmosis, ultrafiltration, microfiltration, activated carbon filters root zone techniques are the tertiary treatment of tannery wastewaters (Labanda et al., 2009; Gallego-Molina et al., 2013; Wang et al., 2011; Suganthi et al., 2013). To treat the wastewater with high loads of salts, crystallization by flat-plate collectors with intensified transpiration and increased evaporation rate has come up as a potential method (Buljan et al., 2001). Lofrano et al. (2008) suggested an effective method of enhanced coagulation for removing the toxic compounds present in the wastewater of leather industries. Subsidies were given to the leather industries and are encouraged by the Government of India to construct Common Effluent Treatment Plants (CETPs) for the treatment of toxic effluents released by the leather manufacturing industries. Some options, such as constructed wetlands (CWs) and integrated wastewater treatment plants, are also encouraged (Tadesse et al., 2004; Calheiros et al., 2008, 2012). However, the problems related to environmental degradation are not resolved because of the sky-high costs which have led to illegal dumping (Beg and Ali, 2008). Additionally, to increase efficiency and cost-effectiveness, substantial developments are being reported at the end-of-pipe treatment in the conventional treatment processes to meet the standards. But the idea of recycling and reusing the tannery liquid discharges after eliminating the toxic contents in leather processing is a better way for wastewater management. It is recommended that preferably near-zero to zero discharge of waste liquors should be appreciated.

5.3.2 Better water management approaches

Of the total water consumption in the leather processing industry, the pre-tanning and tanning processes alone consume about 50%. These processes are estimated to consume about 20 billion liters of water per year in India and around 200 billion liters globally. Therefore, the processes involved in pre-tanning and tanning needs much attention for the development of in-plant measures. Rao et al. (2003) suggest few approaches to overcome the economic problems related to treatment, prevent the wasting of chemicals and water and to reduce pollution. The techniques are as follows:

(i) Recycling and reusing the wastewater
(ii) Revaluating the actual water requirement of tannery operations

These approaches can help in saving water usage during the pre-tanning and tanning processes.

Soaking

Soaking the salted skin/hide by the conventional process consumes 25% of the water when compared to soaking green skin/hide. Therefore, to reduce the water consumption in tannery processes, recycling the soak liquor of salted skin/hide is required. Owing to the presence of soil and dirt in larger amounts, recycling the soak liquor is challenging. The method of counter-current soaking in which, water flows in the opposite direction to the flow of skins/hides can be adopted for efficient recycling (Fig. 3). This method will save 67% of the water utilized for soaking process. By using counter-current soaking, 8 L of water per kilogram of skins/hides was saved in the process where 12 L/kg of skins/hides was consumed. The resulting leathers also exhibited similar physical properties to that of normally processed leathers.

Liming

A large amount of water is used for the liming processes, approximately 4–6 L/kg of skins/hides processed. During this process, water usage can be reduced by recycling the water once used lime liquor for the subsequent batch. In a study, liming liquor was recycled by removing solid matter, adjusting the temperature, and adding new lime-sulfides and water. By this method, lime-sulfide unhairing liquors are recycled for more than 20 times

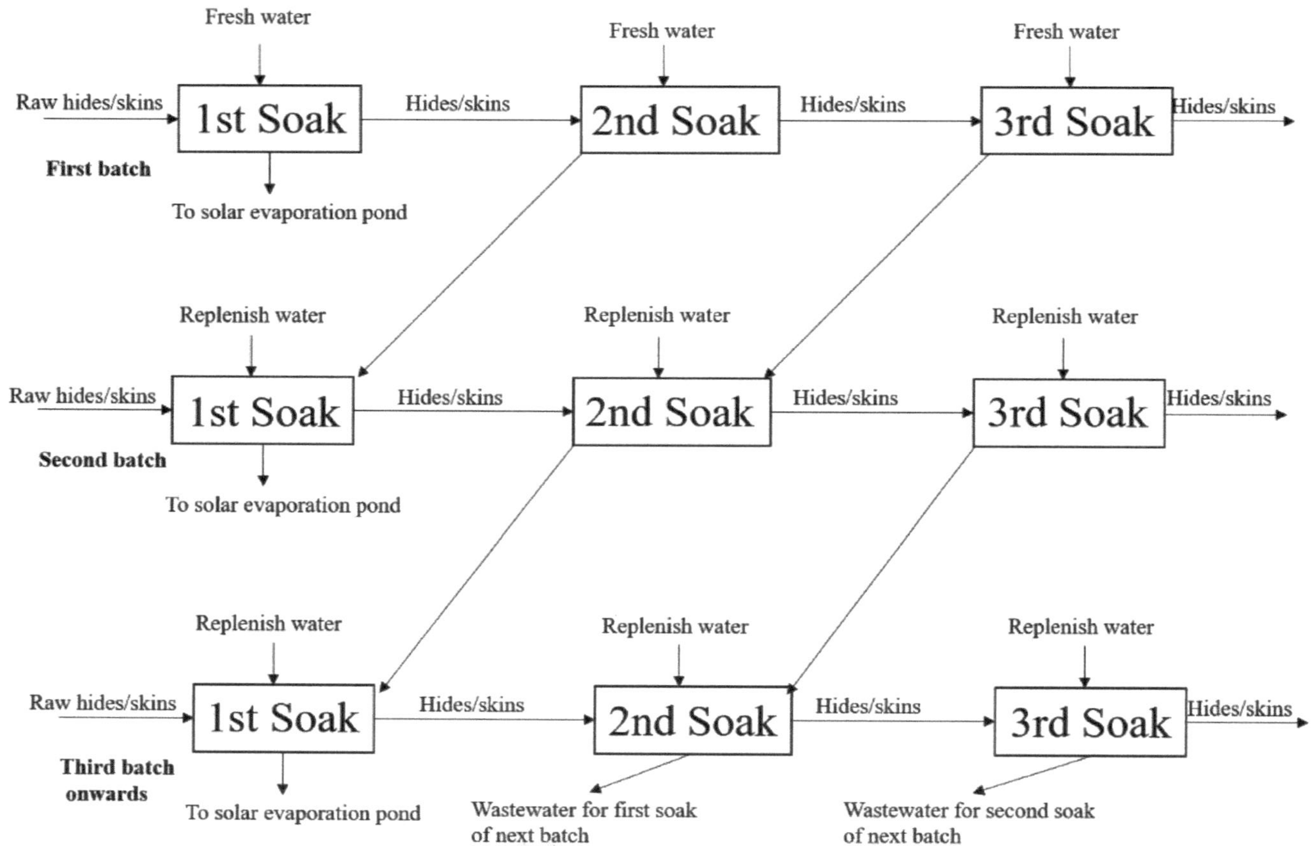

Fig. 3 Schematic representation of counter-current soaking. *(Reprinted from Rao, J.R., Chandrababu, N.K., Muralidharan, C., Nair, B.U., Rao, P.G., Ramasami, T., Recouping the wastewater: a way forward for cleaner leather processing, J. Clean. Prod. 11 (5), 595, Copyright (2002), with permission from Elsevier.)*

saving the water consumption by 20 times (Money and Adminis, 1974). The application of counter-current method can save water usage up to 50%. The physical properties of the obtained leathers were also significant to that of control leathers. As discussed earlier in this chapter, enzymatic unhairing is gaining attention in beamhouse operations (Thanikaivelan et al., 2000). An attempt by Li et al. (2020) to achieve surfactant-free beamhouse operations reduced the water consumption by 20.80% and reduced the COD levels of the spent liquors by 26.90%. Water consumption in this process can be completely reduced by reusing the relimed liquor of skins/hides which are unhaired by enzymatic methods.

Deliming and bating

The effluent characteristics of the tannery effluent are adversely affected by the ammonium salts used during deliming operations. It contributes to about 75–80% of the ammonia content in the effluents (Rao et al., 2003). Usually, water usage is about 1–2 L/kg of skins/hides. Water usage in this operation can be reduced up to 85% by following a floatless operation for deliming and bating or with least float. The resulting leather is of good quality with no characteristic changes in its physical properties (Sundar et al., 2001). Still, due to the presence of large amounts of calcium and ammonium salts, the delimed liquor cannot be recycled.

Pickling and tanning

The conventional pickling process employs 80–100% of water and 8–10% of salt. To initiate chrome tanning the pH is maintained at 2.5–3.0 by adding concentrated sulfuric acid. The release of this liquor will affect the water streams adversely due to the presence of large quantities of neutral salts in pickle liquors. Moreover, a high concentration of sulfates has adverse effects on the process of biological treatment systems (Colleran et al., 1995). The conventional leather processes use about 50% of pickle float as tanning float. The integration of pickling and tanning (closed-loop system) can help in reducing water consumption. To overcome these problems, salt-less pickling, pickle recycle, and pickle-less tanning are some other options. Reusing the pickle liquors for the next lot and adopting pickle-less alum chrome tanning can significantly reduce the TDS levels (Rao et al., 2004; Sivakumar et al., 2005). Another scientific approach is increasing chrome exhaust by employing various tanning auxiliaries as discussed earlier in *High Exhaust Tanning*. This method also helps in reusing the spent liquor in pickling, thereby limiting the levels of pollution in water discharges with added cost-cutting by reducing the amount of chemicals employed during post-tanning processes such as retanning, finishing, and dyeing. By this method, the manufacturers benefit a saving of about 20–40 US$ per metric ton of skins/hides.

Washings

The washing of leather materials and floors during leather processing consumes about 33–38% of total water usage. The inefficient use of water can lead to increase the water consumption up to 50–80 L per kg of skin/hides processed (UNEP, 1991). Proper addition of water to wash the materials and controlled usage for floor washings can reduce the quantity of water used for washings by 40–50%. The continuous water washing system in which the materials are continuously washed in a drum where lattice door is open and water supply is fully open is a major process causing excessive wastage of water. The only way to tremendously reduce the water wastage in this step is by replacing the continuous washing system with a batch system where more than 50% of water wastage can be saved.

6 Conclusion

From the critical analysis of the conventional processes of the leather industry, it is clear that pre-tanning, tanning, and post-tanning through the finishing steps of a tannery are causing environmental deterioration to an extreme level. The first two steps, that is the pre-tanning and tanning, are the major cause for increasing the pollution loads in the effluents of the tanneries. The effluent sludge from the tanneries has severe effects on agriculture and agro-dependent industries by degrading the groundwater quality in the locality. The negative impression of the leather industry is demanding attention toward the need for green technology and effective treatment and reutilization of the waste generated by the industry. Chemical-based process replaced by enzyme actions for major tannery operations, such as soaking, liming, degreasing, dehairing, tanning, and improvement of chrome exhaustion using tanning auxiliaries like biotannin, NPP, NPD, and so on, are cleaner production technologies with environment-friendly methods that greatly reduce the pollution loads Fig. 4. Many studies have proved the potential usage of solid wastes such as fleshing, keratinous materials, wet blues, chrome shaving, trimmings obtained from tannery operations for the development of various products and applications such as in

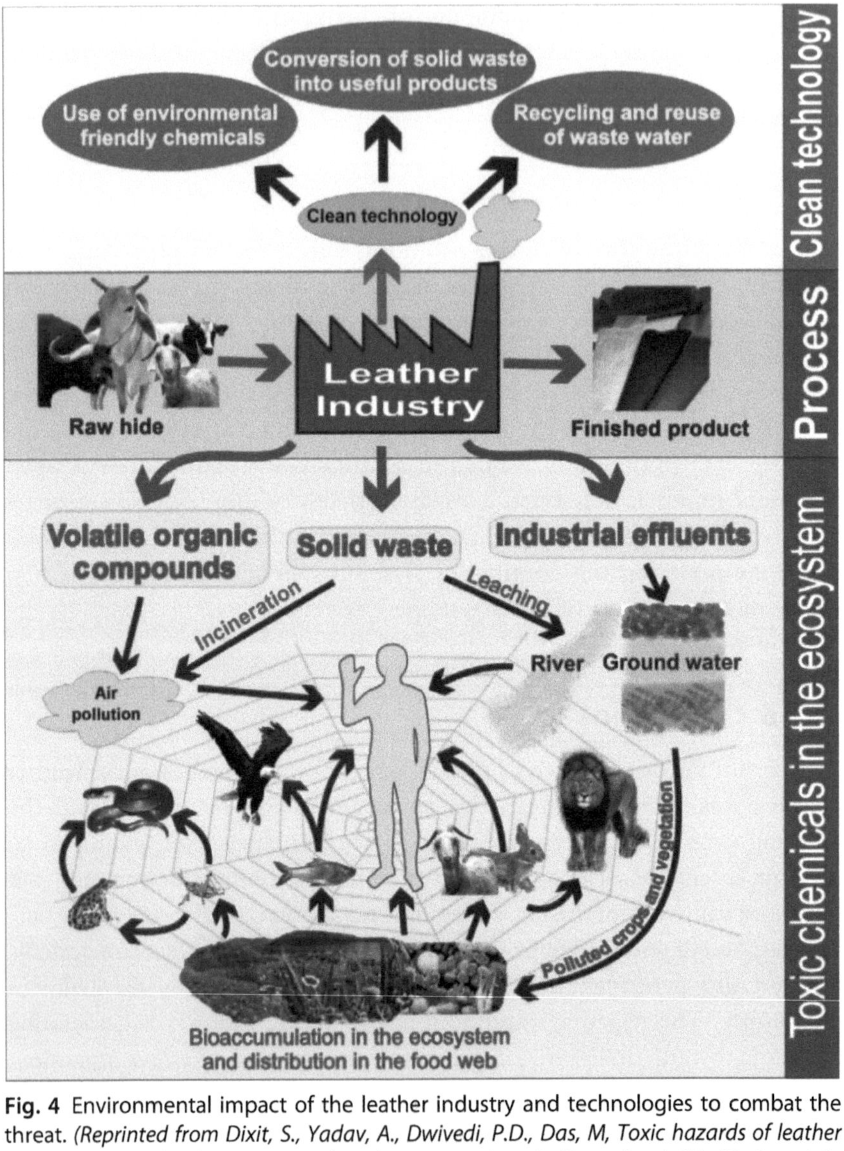

Fig. 4 Environmental impact of the leather industry and technologies to combat the threat. *(Reprinted from Dixit, S., Yadav, A., Dwivedi, P.D., Das, M, Toxic hazards of leather industry and technologies to combat threat: a review, J. Clean. Prod. 87, 40, Copyright (2014), with permission from Elsevier.)*

leather processing, tanning auxiliaries, animal/poultry feed, fertilizers, adhesives, energy generation, cosmetics, biomedical applications, and so on.

The developments in tannery waste treatment technologies, such as biological degradation of sludge and ozonation, are some moves that contribute to reducing the burden of pollution on the environment. Though these processes may are quite expensive they also have several paybacks in relation to conserving the environment, improved labor efficiency, materials with eminent quality, and reliability. The benefits can also be enjoyed by the staff personnel by ensuring the management of occupational hazard and the people living in the same neighborhood. The steps and measures presented in this chapter for adopting cleaner technologies in the leather industry may be quite unbending; howbeit regulators need to follow them for safeguarding the environment and human health. Overall, it is understood that the leather industries are taking various steps and action plans and thus take a giant leap toward production with zero waste.

Important websites

1. https://www.no-burn.org/introduction-to-zero-waste/
2. https://www.clri.org/
3. https://leatherindia.org/indian-leather-industry
4. https://leatherpanel.org/
5. http://www.fao.org/home/en/
6. https://www.unido.org/researchers/publications

References

Afşar, A., Buğra, J., Gülümser, G., Aslan, A., 2010. A study on usability of collagen hydrolysate along with oxazolidine in leather processing. Text. Apparel 20 (1), 37–40.

Aravindhan, R., Saravanabhavan, S., Rao, J.R., Nair, B.U., Thanikaivelan, P., Chandrasekaran, B., 2004. Bio-driven lime and pickle free tanning paves way for greener garment leather production. J. Am. Leather Chem. Assoc. 99 (2), 53–66.

Aravindhan, R., Saravanabhavan, S., Thanikaivelan, P., Rao, J.R., Nair, B.U., 2007. A chemo-enzymatic pathway leads towards zero discharge tanning. J. Clean. Prod. 15 (13–14), 1217–1227.

Asad, S., Amoozegar, M.A., Pourbabaee, A., Sarbolouki, M.N., Dastgheib, S.M.M., 2007. Decolorization of textile azo dyes by newly isolated halophilic and halotolerant bacteria. Bioresour. Technol. 98 (11), 2082–2088.

Aslan, A., Gueluemser, G., January, B., 2006. Increased chromium tanning efficiency with collagen hydrolysates. J. Soc. Leather Technol. Chem. 90, 201–204.

Aslan, A.H.M.E.T., Karavana, H.A., Gulumser, G., Yasa, I., Cadirci, B.H., 2007. Utilization of collagen hydrolysate in keratinase production from Bacillus subtilis ATCC 6633. Ile 1 (1.2), 1–4.

Bacardit, A., van der Burgh, S., Armengol, J., Ollé, L., 2014. Evaluation of a new environment friendly tanning process. J. Clean. Prod. 65, 568–573.

Beg, K.R., Ali, S., 2008. Chemical contaminants and toxicity of ganga river sediment from up and down stream area at Kanpur. Am. J. Environ. Sci. 4 (4), 362.

Bensouici, F., Tala-Ighil, R., Bououdina, M., 2016. Nanostructured oxide materials for photodegradation of dyes. In: Advanced Environmental Analysis, pp. 207–234.

Bhaskar, N., Sakhare, P.Z., Suresh, P.V., Gowda, L.R., Mahendrakar, N.S., 2007. Biostabilization and preparation of protein hydrolysates from delimed leather fleshings. J. Sci. Ind. Res. 66, 1054–1063.
Brown, E.M., Taylor, M.M., Marmer, W.N., 1996. Production and potential uses of co-products from solid tannery waste. J. Am. Leather Chem. Assoc. 91 (10), 270–276.
Buljan, J., Sahasranaman, A., Sampathkumar, S., Mani, A., Venkatesh, A., 2001. Accelerated evaporation of saline streams in solar pans. In: Proceedings of the XXVI IULTCS Congress, Cape Town.
Cabeza, L.F., Taylor, M.M., DiMaio, G.L., Brown, E.M., Marmer, W.N., Carrio, R., Celma, P.J., Cot, J., 1998. Processing of leather waste: pilot scale studies on chrome shavings. Isolation of potentially valuable protein products and chromium. Waste Manag. 18 (3), 211–218.
Cakl, J., Medved, Å.Ã.J., ÃovÃ, L., Blaha, A., 1998. Demineralization of protein hydrolysates from enzymatic hydrolysis of leather shavings using membrane diafiltration. Sep. Sci. Technol. 33 (9), 1271–1281.
Calheiros, C.S., Rangel, A.O., Castro, P.M., 2008. Evaluation of different substrates to support the growth of Typha latifolia in constructed wetlands treating tannery wastewater over long-term operation. Bioresour. Technol. 99 (15), 6866–6877.
Calheiros, C.S., Quitério, P.V., Silva, G., Crispim, L.F., Brix, H., Moura, S.C., Castro, P. M., 2012. Use of constructed wetland systems with Arundo and Sarcocornia for polishing high salinity tannery wastewater. J. Environ. Manag. 95 (1), 66–71.
Cantera, C.S., 2003. Adding value to chrome shavings: hydrolysates as retanning materials. World Leath. 16, 27–29.
Cantera, C., Martegani, J., Esterelles, G., Vergara, J., 2002. Collagen hydrolysate: 'soluble skin' applied in post-tanning processes. Part 2: Interaction with acrylic retanning agents. J. Soc. Leath. Tech. Chem. 86 (5), 195–202.
Castiello, D., Puccini, M., Salvadori, M., Vitolo, S., 2006. Reutilization of skin fleshing-derived collagen hydrolizate in the re-tanning/dyeing/fatliquoring phases. In: IULTCS II Euro Congress. [online] Instanbul. Available at: http://www.polotecnologico.com/file/reutilizationofskinfleshing-derivedcollagenhydrolizateintheretanningdyeingfatliquoring.pdf. (Accessed 24 April 2020).
Castilhos, D.D., Tedesco, M.J., Vidor, C., 2002. Crop yields and chemical changes in the soil treated with tannery residues and hexavalent chromium. Brazil. J. Soil Sci. 26 (4), 1083–1092.
Catalina, M., Antunes, A.P.M., Attenburrow, G., Cot, J., Covington, A.D., Phillips, P.S., 2007. Sustainable management of waste-reduction of the chromium content of tannery solid waste as a step in the cleaner production of gelatin. J. Solid Waste Technol. Manag. 33 (1), 43–50.
Cavello, I.A., Cavalitto, S.F., 2013. Optimization of Keratinolytic Protease Production by Purpureocillium Lilacinum LPS# 876 as a Sustainable Management of Tannery Hair Waste.
Chakraborty, R., 2003. Ecofriendly solid waste management in leather industry. Environ. Biotechnol. Newslett. (University of Kalyani) 3 (2).
Colak, S.E.L.I.M.E., Zengin, G.Ö.K.H.A.N., Ozgunay, H., Sarikahya, H., Sari, O., Yuceer, L., 2005. Utilization of leather industry prefleshings in biodiesel production. J. Am. Leather Chem. Assoc. 100 (4), 137–141.
Colicchia, C., Melacini, M., Perotti, S., 2011. Benchmarking supply chain sustainability: insights from a field study. BIJ 18 (5), 705–732.
Colleran, E., Finnegan, S., Lens, P., 1995. Anaerobic treatment of sulphate-containing waste streams. Antonie Van Leeuwenhoek 67 (1), 29–46.
Cooman, K., Gajardo, M., Nieto, J., Bornhardt, C., Vidal, G., 2003. Tannery wastewater characterization and toxicity effects on Daphnia spp. Environ. Toxicol. 18 (1), 45–51.

Cot, J., Manich, A.M., Marsal, A., Fort, M., Celma, P.J., Carrio, R., Choque, R., Cabeza, L. F., 1999. Processing of collagenic residues. Isolation of gelatin by the action of peroxochromates. J. Am. Leather Chem. Assoc. 94 (4), 115–128.
Council for Leather Exports, 2020. Indian Leather Industry – Overview, Export Performance & Prospects. [online]. Available at: https://leatherindia.org/indian-leather-industry/. (Accessed 12 April 2020).
Crispim, A., Mota, M., 2003. Leather shavings treatment-an enzymatic approach. J. Soc. Leather Technol. Chem. 87, 203–207.
CSIRO, 2001. Salinity Reduction in Tannery Effluents in India & Australia. Project Proposal to ACIAR by CSIRO Textile & Fibre Technology, Leather Research Centre.
Dayanandan, A., Kanagaraj, J., Sounderraj, L., Govindaraju, R., Rajkumar, G.S., 2003. Application of an alkaline protease in leather processing: an ecofriendly approach. J. Clean. Prod. 11 (5), 533–536.
De Aquim, P.M., Gutterres, M., Trierweiler, J., 2010. Assessment of water management in tanneries. J. Soc. Leather Technol. Chem. 94 (6), 253–258.
Deng, H., 2015. Multicriteria analysis for benchmarking sustainability development. BIJ 22 (5), 791–807.
Dettmer, A., Coelho Cavalheiro, J., Cavalli, E., Misturini Rossi, D., de Souza Gusatti, C., Zachia Ayub, M.A., Gutterres, M., 2012. Optimization of the biotechnological process for hide unhairing in substitution of toxic sulfides. Chem. Eng. Technol. 35 (5), 803–810.
Dhayalan, K., Fathima, N.N., Gnanamani, A., Rao, J.R., Nair, B.U., Ramasami, T., 2007. Biodegradability of leathers through anaerobic pathway. Waste Manag. 27 (6), 760–767.
Dixit, S., Yadav, A., Dwivedi, P.D., Das, M., 2015. Toxic hazards of leather industry and technologies to combat threat: a review. J. Clean. Prod. 87, 39–49.
Dowlath, M., Karuppannan, S., Rajadesingu, S., Arunachalam, K., 2020. Major environmental issues and problems. In: Hussain, C. (Ed.), Handbook of Environmental Materials Management. Springer, Cham.
Durga, J., Ranjithkumar, A., Ramesh, R., Girivasan, K.T.P.V., Rose, C., Muralidharan, C., 2016. Replacement of lime with carbohydrases–a successful cleaner process for leather making. J. Clean. Prod. 112, 1122–1127.
Dvorackova, M., Julinova, M., Mikulcakova, H., Kupec, J., 2007. Testing aerobic degradation of the protein portion of filter cake after filtration of chrome tanned waste. J. Am. Leather Chem. Assoc. 102 (3), 77–84.
El Boushy, A.R., Van Der Poel, A.F.B., Koene, J.I.A., Dieleman, S.H., 1991. Tanning waste by-product from cattle hides, its suitability as a feedstuff. Bioresour. Technol. 35 (3), 321–323.
El Samra, G.H., Hussein, A.A., Farahat, S.A., Samir, A.M., 2007. Lipid peroxidation due to tri-valent chromium (CR3) among tannery workers. Egypt. J. Occup. Med. 31 (1), 87–105.
Elisangela, F., Andrea, Z., Fabio, D.G., de Menezes Cristiano, R., Regina, D.L., Artur, C.P., 2009. Biodegradation of textile azo dyes by a facultative Staphylococcus arlettae strain VN-11 using a sequential microaerophilic/aerobic process. Int. Biodeterior. Biodegradation 63 (3), 280–288.
El-Sheikh, M.A., Saleh, H.I., Flora, J.R., AbdEl-Ghany, M.R., 2011. Biological tannery wastewater treatment using two stage UASB reactors. Desalination 276 (1–3), 253–259.
Environmental Protection Agency, 1982. Development Document for Effluent Limitations Guidelines New Source Performance Standards and Pretreatment Standards for the Leather Tanning and Finishing Point Source Category. EPA-440/1-82-016, Effluent Guidelines Division, Office of Water, U.S. Environmental Protection Agency.
Famielec, S., Wieczorek-Ciurowa, K., 2011. Waste from leather industry. Threats to the environment. Czasopismo Techniczne. Chemia 108 (1-Ch), 43–48.

Farabegoli, G., Carucci, A., Majone, M., Rolle, E., 2004. Biological treatment of tannery wastewater in the presence of chromium. J. Environ. Manag. 71 (4), 345–349.
Ferreira, M.J., Almeida, M.F., 2011. Recycling of leather waste containing chromium-a review. Mater. Sci. Res. J. 5 (4), 327.
Ferreira, M.J., Almeida, M.F., Pinho, S.C., Gomes, J.R., Rodrigues, J.L., 2014. Alkaline hydrolysis of chromium tanned leather scrap fibers and anaerobic biodegradation of the products. Waste Biomass Valoriz. 5 (4), 551–562.
Food and Agriculture Organization of the United Nations, 2016. World Statistical Compendium for Raw Hides and Skins, Leather and Leather Footwear 1999–2015. [online]. Food and Agriculture Organization of the United Nations, Rome. Available at http://www.fao.org/publications/card/en/c/2bb7ce63-da1b-4d89-9510-afc725b5e960/. (Accessed 12 April 2020).
Frendrup, W., 1995. UNEP Cleaner Production Industrial Sector Guide. Leather Industry, Taastrup.
Gaidau, C., Ghiga, M., Filipescu, L., Stepan, E., Lacatus, V., Popescu, M., 2007, October. Advanced materials obtained from leather–by products. In: CEEX Conference Proceedings Romania, p. 252.
Gaidău, C., Ghiga, M., Stepan, E., Taloi, D., Filipescu, L., 2009. Additives and advanced biomaterials obtained from leather industry by-products. Rev. Chim. (Bucharest, Rom.) 60 (5), 501–507.
Gallego-Molina, A., Mendoza-Roca, J.A., Aguado, D., Galiana-Aleixandre, M.V., 2013. Reducing pollution from the deliming–bating operation in a tannery. Wastewater reuse by microfiltration membranes. Chem. Eng. Res. Des. 91 (2), 369–376.
Ganesh, R., Balaji, G., Ramanujam, R.A., 2006. Biodegradation of tannery wastewater using sequencing batch reactor—respirometric assessment. Bioresour. Technol. 97 (15), 1815–1821.
Genovese, A., Lenny Koh, S.C., Kumar, N., Tripathi, P.K., 2014. Exploring the challenges in implementing supplier environmental performance measurement models: a case study. Prod. Plan. Control 25 (13–14), 1198–1211.
Gish, A.J., 2000. Leatherboard, a practical use of tannery offal. J. Am. Leath. Chem. Assoc. 95 (2), 43–47.
Global Alliance for Incinerator Alternatives, 2020. Introduction to Zero Waste - Global Alliance for Incinerator Alternatives. [online]. Available at: https://www.no-burn.org/introduction-to-zero-waste/. (Accessed 25 April 2020).
Gnanamani, A., Bai, R.K., 1992. Influence of biodigested slurry on rice-gram cultivation. Bioresour. Technol. 41 (3), 217–221.
Guerrero, L.A., Maas, G., Hogland, W., 2013. Solid waste management challenges for cities in developing countries. Waste Manag. 33 (1), 220–232.
Gupta, S., Babu, B.V., 2009. Removal of toxic metal Cr (VI) from aqueous solutions using sawdust as adsorbent: equilibrium, kinetics and regeneration studies. Chem. Eng. J. 150 (2–3), 352–365.
Hagerman, A.E., 1980. Ph.D. Thesis. Purdue University, West Lafayette, IN.
Han, B., Jaurequi, J., Tang, B.W., Nimni, M.E., 2003. Proanthocyanidin: a natural crosslinking reagent for stabilizing collagen matrices. J. Biomed. Mater. Res. A 65 (1), 118–124.
He, F., Hu, W., Li, Y., 2004. Biodegradation mechanisms and kinetics of azo dye 4BS by a microbial consortium. Chemosphere 57 (4), 293–301.
Hosseinpour, A., Peng, Q., Gu, P., 2015. A benchmark-based method for sustainable product design. BIJ 22 (4), 643–664.
Hrncirik, J., Dvorackova, M., Hruska, F., Kupec, J., Kolomaznik, K., 2005. Kinetics of alkaline hydrolysis of the insoluble protein fraction following enzymatic hydrolysis of chrome-tanned shavings. J. Am. Leather Chem. Assoc. 100 (1), 1–7.

Hu, J., Xiao, Z., Zhou, R., Deng, W., Wang, M., Ma, S., 2011. Ecological utilization of leather tannery waste with circular economy model. J. Clean. Prod. 19 (2–3), 221–228.
Hussain, C.M. (Ed.), 2019. Handbook of Environmental Materials Management. Springer, Cham, Switzerland.
Hussain, C.M., Kecili, R., 2019. Modern Environmental Analysis Techniques for Pollutants. Elsevier.
Hussain, C., Kharisov, B., 2017. Advanced Environmental Analysis: Applications of Nanomaterials. Royal Society of Chemistry.
Hussain, C.M., Mishra, A.K., 2018. Nanotechnology in Environmental Science, 2 Volumes. Vol. 1 John Wiley & Sons.
Hussain, C.M., Solcova, O., Kacem-Boureau, M., dos Santos, E.V., Stegemann, J., Alvarez, E.T., Karima, G., Cristaldi, A., Kuppusamy, S., Castro, P.M.L., Castelo-Grande, T., 2020. The Handbook of Environmental Remediation: Classic and Modern Techniques. Royal Society of Chemistry.
International Union of Environment Commission, 2018. IUE 2 - Recommendations for Tannery Solid By-Product Management. [online]. International Union of Environment Commission (IUE), London. Available at: http://www.iultcs.org/environment-iue.php. (Accessed 24 April 2020).
Jayanthi, D., Victor, J.S., Chellan, R., Chellappa, M., 2019. Green processing: minimising harmful substances in leather making. Environ. Sci. Pollut. Res. 26 (7), 6782–6790.
Kamini, N.R., Hemachander, C., Mala, J.G.S., Puvanakrishnan, R., 1999. Microbial enzyme technology as an alternative to conventional chemicals in leather industry. Curr. Sci., 80–86.
Kanagaraj, J., Mandal, A.B., 2012. Combined biodegradation and ozonation for removal of tannins and dyes for the reduction of pollution loads. Environ. Sci. Pollut. Res. 19 (1), 42–52.
Kanagaraj, J., Senthilvelan, T., Panda, R.C., Kavitha, S., 2015. Eco-friendly waste management strategies for greener environment towards sustainable development in leather industry: a comprehensive review. J. Clean. Prod. 89, 1–17.
Kanagaraj, J., Panda, R.C., Senthilvelan, T., Gupta, S., 2016. Cleaner approach in leather dyeing using graft copolymer as high performance auxiliary: related kinetics and mechanism. J. Clean. Prod. 112, 4863–4878.
Karabay, S., 2008. Waste Management in Leather Industry. Postgraduate, Dokuz Eylul University.
Karthikeyan, R., Balaji, S., Sehgal, P.K., 2007. Industrial applications of keratins—a review. J. Sci. Ind. Res. 66, 710–715.
Karupannan, S.K., Dowlath, M.J.H., Arunachalam, K.D., 2020. Phytonanotechnology: challenges and future perspectives. Phytonanotechnol. Chall. Prosp., 303.
Katsifas, E.A., Giannoutsou, E., Lambraki, M., Barla, M., Karagouni, A.D., 2004. Chromium recycling of tannery waste through microbial fermentation. J. Ind. Microbiol. Biotechnol. 31 (2), 57–62.
Kaul, S.N., NAndy, T., Vyas, R.D., Szpyrkowicz, L., 2001. Waste management in tanneries: experience and outlook. J. Indian Assoc. Environ. Manag. 28, 56–76.
Kolomaznik, K., Uhlířová, M., Fürst, T., 2009. Possibility of using tannery waste for biodiesel production. J. Am. Leather Chem. Assoc. 104, 177–182.
Konrad, E.E., Castilhos, D.D., 2002. Chemical changes in soil and corn growth resulting from the addition of tannery sludge. Brazil. J. Soil Sci. 26 (1), 257–265.
Krishnamoorthy, G., Sadulla, S., Sehgal, P.K., Mandal, A.B., 2012. Green chemistry approaches to leather tanning process for making chrome-free leather by unnatural amino acids. J. Hazard. Mater. 215, 173–182.
Kumar, A.G., Swarnalatha, S., Sairam, B., Sekaran, G., 2008. Production of alkaline protease by Pseudomonas aeruginosa using proteinaceous solid waste generated from leather manufacturing industries. Bioresour. Technol. 99 (6), 1939–1944.

Kupec, J., Dvorackava, M., Rudlova, S., Ruzicka, J., Kolomaznik, K., 2002. Deproteination of chrome waste by washing and enzymatic hydrolysis. J. Am. Leather Chem. Assoc. 97 (9), 349–354.

Labanda, J., Khaidar, M.S., Llorens, J., 2009. Feasibility study on the recovery of chromium (III) by polymer enhanced ultrafiltration. Desalination 249 (2), 577–581.

Lefebvre, O., Vasudevan, N., Torrijos, M., Thanasekaran, K., Moletta, R., 2006. Anaerobic digestion of tannery soak liquor with an aerobic post-treatment. Water Res. 40 (7), 1492–1500.

Li, S., Luo, F., Chattha, S.A., Zhang, C., Peng, B., Mu, C., 2020. Surfactant-free beamhouse technology of leather manufacturing: removing constraints for the breakdown of natural fats catalyzed by lipase. J. Clean. Prod., 121187.

Lima, D.Q., Oliveira, L.C.A., Bastos, A.R.R., Carvalho, G.S., Marques, J.G.S.M., Carvalho, J.G., De Souza, G.A., 2010. Leather industry solid waste as nitrogen source for growth of common bean plants. Appl. Environ. Soil Sci. 2010.

Liu, Y., Yang, Z., 2006. Uses for chrome leather waste-chromium enriched brewers yeast. J. Soc. Leather Technol. Chem. 90 (1), 14.

Lofrano, G., Aydin, E., Russo, F., Guida, M., Belgiorno, V., Meric, S., 2008. Characterization, fluxes and toxicity of leather tanning bath chemicals in a large tanning district area (IT). Water, Air, & Soil Pollution: Focus 8 (5–6), 529–542.

Ma, J., Hou, X., Gao, D., Lv, B., Zhang, J., 2014. Greener approach to efficient leather soaking process: role of enzymes and their synergistic effect. J. Clean. Prod. 78, 226–232.

Malek, A., Hachemi, M., Didier, V., 2009. New approach of depollution of solid chromium leather waste by the use of organic chelates: economical and environmental impacts. J. Hazard. Mater. 170 (1), 156–162.

Minghua, Z., Xiumin, F., Rovetta, A., Qichang, H., Vicentini, F., Bingkai, L., Giusti, A., Yi, L., 2009. Municipal solid waste management in Pudong new area, China. Waste Manag. 29 (3), 1227–1233.

Money, C.A., Adminis, U., 1974. Recycling of lime-sulphide unhairing liquors. I. Small-scale trials. J. Soc. Leath. Tech. Chem. 58, 35–40.

Montoneri, E., Rizzi, G., Rizzi, A., Mordenti, A., Bauli, A., Riolfatti, M., Pellegrini, L., 1994. Hydrolysis of tannery wastes to protein meal for animal feedstuffs: a process and product evaluation. J. Chem. Technol. Biotechnol.: Int. Res. Process, Environ. Clean Technol. 59 (1), 91–99.

Morera, J.M., Bacardit, A., Olle, L., Costa, J., Germann, H.P., 2006. Study of a chrome tanning process without float and with low-salt content as compared to a traditional process. II. J. Am. Leather Chem. Assoc. 101 (12), 454–460.

Morera, J.M., Bartolí, E., Chico, R., Solé, C., Cabeza, L.F., 2011. Minimization of the environmental impact of chrome tanning: a new process reusing the tanning floats. J. Clean. Prod. 19 (17–18), 2128–2132.

Morera, J.M., Bartolí, E., Singla, C., 2013. Effect of ultrasound on bovine and ovine skins soaking. J. Clean. Prod. 59, 79–85.

Mu, C., Lin, W., Zhang, M., Zhu, Q., 2003. Towards zero discharge of chromium-containing leather waste through improved alkali hydrolysis. Waste Manag. 23 (9), 835–843.

Munoz, J., Maldonado, M., Rangel, A., 2002. Development of a tanning process based on using hydrolyzated material collected from leather scrap. J. Am. Leather Chem. Assoc. 97 (2), 83–88.

Mwinyihija, M., 2010. Ecotoxicological Diagnosis in the Tanning Industry. Springer Science & Business Media.

Mwinyihija, M., 2012. Pollution control and remediation of the tanning effluent. Open Environ. Pollut. Toxicol. J. 3 (1).

Nashy, E., Eid, K.A., 2019. High exhaustion of chrome tan, enhancement of leather properties and reduction of chrome tanning effluent impact. Egypt. J. Chem. 62 (3), 415–428.
Nazer, D.W., Siebel, M.A., 2006. Reducing the environmental impact of the unhairing–liming process in the leather tanning industry. J. Clean. Prod. 14 (1), 65–74.
Ngoc, U.N., Schnitzer, H., 2009. Sustainable solutions for solid waste management in southeast Asian countries. Waste Manag. 29 (6), 1982–1995.
Ozgunay, H., Colak, S., Mutlu, M.M., Akyuz, F., 2007. Characterization of leather industry wastes. Pol. J. Environ. Stud. 16 (6).
Pati, A., Chaudhary, R., Subramani, S., 2014. A review on management of chrome-tanned leather shavings: a holistic paradigm to combat the environmental issues. Environ. Sci. Pollut. Res. 21, 11266–11282.
Paul, H., Antunes, A.P.M., Covington, A.D., Evans, P., Phillips, P.S., 2013. Towards zero solid waste: utilising tannery waste as a protein source for poultry feed. The 28th International Conference on Solid Waste Technology and Management. The Journal of Solid Waste Technology and Management, Philadelphia, USA.
Pfleiderer, E., 1983. Optimale nassentfettung von rohhaut, blössen und wetblues. Leder 34, 181–186.
Rai, D., Eary, L.E., Zachara, J.M., 1989. Environmental chemistry of chromium. Sci. Total Environ. 86 (1–2), 15–23.
Rajamani, S., Chen, Z., Zhang, S., Su, C., 2009. Recent developments in cleaner production and environment protection in world leather sector. In: 30th IULTCS Congress, p. 5.
Rao, J.R., Thanikaivelan, P., Sreeram, K.J., Nair, B.U., 2002b. Green route for the utilization of chrome shavings (chromium-containing solid waste) in tanning industry. Environ. Sci. Technol. 36 (6), 1372–1376.
Rao, J.R., Chandrababu, N.K., Muralidharan, C., Nair, B.U., Rao, P.G., Ramasami, T., 2003. Recouping the wastewater: a way forward for cleaner leather processing. J. Clean. Prod. 11 (5), 591–599.
Rao, J.R., Chandrasekaran, B., Nair, B.U., Ramasami, T., 2002. Eco-benign management options for cleaner chrome tanning. J. Sci. Ind. Res. 61, 912–926.
Rao, J.R., Kanthimathi, M., Thanikaivelan, P., Sreeram, K.J., Ramesh, R., Ramalingam, S., Chandrababu, N.K., Nair, B.U., Ramasami, T., 2004. Pickle-free chrome tanning using a polymeric synthetic tanning agent for cleaner leather processing. Clean Techn. Environ. Policy 6 (4), 243–249.
Ravindranath, E., Gopalakrishnan, A.N., 2010. Enhancement of biomethanization by pretreatment of limed fleshings from tanneries. J. Sci. Ind. Res. 69, 711–716.
Saha, N., Křesálková, M., Saha, L., Kolomaznik, K., 2003. Effect of anaerobic digestion on chrome sludge–a by-product of tanned lather waste. J. Am. Leather Chem. Assoc. 98 (7), 256–262.
Saravanabhavan, S., Aravindhan, R., Thanikaivelan, P., Chandrasekaran, B., Raghava RaO, J., Unni Nair, B., 2003a. An integrated eco-friendly tanning method for the manufacture of upper leathers from goatskins. J. Soc. Leather Technol. Chem. 87 (4), 149–158.
Saravanabhavan, S., Aravindhan, R., Thanikaivelan, P., Rao, J.R., Nair, B.U., 2003b. Green solution for tannery pollution: effect of enzyme based lime-free unhairing and fibre opening in combination with pickle-free chrome tanning. Green Chem. 5 (6), 707–714.
Saravanabhavan, S., Thanikaivelan, P., Rao, J.R., Nair, B.U., Ramasami, T., 2004. Natural leathers from natural materials: progressing toward a new arena in leather processing. Environ. Sci. Technol. 38 (3), 871–879.
Saravanan, P., Renitha, T.S., Gowthaman, M.K., Kamini, N.R., 2014. Understanding the chemical free enzyme based cleaner unhairing process in leather manufacturing. J. Clean. Prod. 79, 258–264.

Sarkis, J., Schneider, J.L., Wilson, A., Rosenbeck, J.M., 2010. Pharmaceutical companies and sustainability: an analysis of corporate reporting. BIJ 17 (3), 421–434.

Savino, M.M., Manzini, R., Mazza, A., 2015. Environmental and economic assessment of fresh fruit supply chain through value chain analysis. A case study in chestnuts industry. Prod. Plan. Control 26 (1), 1–18.

Sekaran, G., Swarnalatha, S., Srinivasulu, T., 2007. Solid waste management in leather sector. J. Des. Manuf. Technol. 01 (01), 47–52. [online] Available at: http://journals-sathyabama.com/archives/dmt/DMT-Int-Journal-Oct-2007-8.pdf. (Accessed 12 April 2020).

Senthilvelan, T., Kanagaraj, J., Mandal, A.B., 2012. Application of enzymes for dehairing of skins: cleaner leather processing. Clean Techn. Environ. Policy 14 (5), 889–897.

Shamma, H., Hassan, S., 2013. Customer-driven benchmarking. BIJ 20 (3), 377–395.

Shin, S.B., Min, B.W., Yang, S.H., Park, M.S., Kim, H.S., Kim, B.H., Baik, D.H., 2008. Production of biodiesel from fleshing scrap using immobilized lipase-catalyst. Appl. Biol. Chem. 51 (3), 177–182.

Silva, R.C., Resende Júnior, J.C.D., Lima, R.F.D., Sousa, R.V.D., Oliveira, L.C.A.D., Daniel, J.L.P., Moreira, A.D.O., 2012. Potential of wet blue leather waste for ruminant feeding. Rev. Bras. Zootec. 41 (4), 1070–1073.

Sivakumar, V., Sundar, V.J., Rangasamy, T., Muralidharan, C., Swaminathan, G., 2005. Management of total dissolved solids in tanning process through improved techniques. J. Clean. Prod. 13 (7), 699–703.

Sivaparvathi, M., Suseela, K., Nanda, S.C., 1986. Hydrolytic action of Pseudomonas aeruginosa on chrome shavings. Leather Sci. 33, 8–11.

Song, Q., Li, J., Zeng, X., 2015. Minimizing the increasing solid waste through zero waste strategy. J. Clean. Prod. 104, 199–210.

Suganthi, V., Mahalakshmi, M., Balasubramanian, N., 2013. Development of hybrid membrane bioreactor for tannery effluent treatment. Desalination 309, 231–236.

Sundar, V.J., Ramesh, R., Rao, P.S., Saravanan, P., Sridharnath, B., Muralidharan, C., 2001. Water management in leather industry. J. Sci. Ind. Res. 60, 443–450.

Sundar, V.J., Raghavarao, J., Muralidharan, C., Mandal, A.B., 2011a. Recovery and utilization of chromium-tanned proteinous wastes of leather making: A review. Crit. Rev. Environ. Sci. Technol. 41 (22), 2048–2075.

Sundar, V.J., Gnanamani, A., Muralidharan, C., Chandrababu, N.K., Mandal, A.B., 2011b. Recovery and utilization of proteinous wastes of leather making: a review. Rev. Environ. Sci. Biotechnol. 10 (2), 151–163.

Sundarapandiyan, S., Chandrasekar, R., Ramanaiah, B., Krishnan, S., Saravanan, P., 2010. Electrochemical oxidation and reuse of tannery saline wastewater. J. Hazard. Mater. 180 (1–3), 197–203.

Sundarapandiyan, S., Brutto, P.E., Siddhartha, G., Ramesh, R., Ramanaiah, B., Saravanan, P., Mandal, A.B., 2011. Enhancement of chromium uptake in tanning using oxazolidine. J. Hazard. Mater. 190 (1–3), 802–809.

Suresh, V., Kanthimathi, M., Thanikaivelan, P., Rao, J.R., Nair, B.U., 2001. An improved product-process for cleaner chrome tanning in leather processing. J. Clean. Prod. 9 (6), 483–491.

Swarnalatha, S., Ganesh Kumar, A., Tandaiah, S., Sekaran, G., 2009. Efficient and safe disposal of chrome shavings discharged from leather industry using thermal combustion. J. Chem. Technol. Biotechnol.: Int. Res. Process, Environ. Clean Technol. 84 (5), 751–760.

Tadesse, I., Green, F.B., Puhakka, J.A., 2004. Seasonal and diurnal variations of temperature, pH and dissolved oxygen in advanced integrated wastewater pond system® treating tannery effluent. Water Res. 38 (3), 645–654.

Taylor, M.M., Cabeza, L.F., Marmer, W.N., Brown, E.M., 1997. Computer-assisted method to measure the adhesive properties of hydrolysis products from collagen. J. Am. Leather Chem. Assoc. 92 (2), 28–37.

Taylor, M.M., Cabeza, L.F., DiMaio, G.L., Brown, E., Marmer, W.N., Carrió, R., Celma, P.J., Cot, J., 1998. Processing of leather waste: pilot scale studies on chrome shavings. Part I. Isolation and characterization of protein products and separation of chrome cake. J. Am. Leather Chem. Assoc. 93 (3), 61–82.

Taylor, M.M., Liu, C.K., Marmer, W.N., Brown, E.M., 2003. Enzymatic modification of hydrolysis products from collagen using a microbial transglutaminase. III. Preparation of films with improved mechanical properties. J. Am. Leather Chem. Assoc. 98 (11), 435–444.

Thanikaivelan, P., Rao, J.R., Nair, B.U., 2000. Development of a leather processing method in narrow pH profile: part 1. Standardisation of dehairing process. J. Soc. Leather Technol. Chem. 84 (6), 276–284.

Thanikaivelan, P., Rao, J.R., Nair, B.U., Ramasami, T., 2003. Biointervention makes leather processing greener: an integrated cleansing and tanning system. Environ. Sci. Technol. 37 (11), 2609–2617.

Thanikaivelan, P., Rao, J.R., Nair, B.U., Ramasami, T., 2004. Progress and recent trends in biotechnological methods for leather processing. Trends Biotechnol. 22 (4), 181–188.

Thanikaivelan, P., Rao, J.R., Nair, B.U., Ramasami, T., 2005. Recent trends in leather making: processes, problems, and pathways. Crit. Rev. Environ. Sci. Technol. 35 (1), 37–79.

Tünay, O., Kabdasli, I., Orhon, D., Ates, E., 1995. Characterization and pollution profile of leather tanning industry in Turkey. Water Sci. Technol. 32 (12), 1–9.

Ueda, K., Takenaka, T., Váncza, J., Monostori, L., 2009. Value creation and decision-making in sustainable society. CIRP Ann. 58 (2), 681–700.

UNEP Industry and Environment Programme Activity Centre, 1991. Tanneries and the Environment: A Technical Guide to Reducing the Environmental Impact of Tannery Operations. Technical report series/WHO, UNEP Industry and Environment Programme Activity Centre, Paris, France, pp. 89–93.

UNIDO, 2010. Future Trends in the World Leather and Leather Products Industry and Trade (2010). [online]. United Nations Industrial Development Organization, Vienna. Available at: https://leatherpanel.org/content/future-trends-and-expected-status-world-leather-and-leather-products-industry-and-trade-2010. (Accessed 12 April 2020).

UNIDO, 2011. Introduction to Treatment of Tannery Effluents [online]. United Nations Industrial Development Organization, Vienna. Available at: https://www.unido.org/resources/publications/creating-shared-prosperity/agribusiness-and-rural-entrepreneurship/introduction-treatment-tannery-effluents (Accessed 25 April 2020).

Wang, H., Wang, Y., Zhou, L., 2011, June. Purification and recycling of tannery degreasing wastewater by ultrafiltration with polyimide membrane. In: 2011 International Conference on Remote Sensing, Environment and Transportation Engineering. IEEE, pp. 569–572.

Zhang, X., Zhang, Y., 1982. Test use of alkaline lipase in degreasing of pigskin. Pige Keji 40, 16.

CHAPTER EIGHT

Modern society and zero waste tools

Suriyaprakash Rajadesingu, V. Deepankara, Mohammed Junaid Hussain Dowlath, Sathish Kumar Karuppannan, and Kantha Deivi Arunachalam
Center for Environmental Nuclear Research, Directorate of Research, SRM Institute of Science and Technology, Kattankulathur, Tamil Nadu, India

Contents

1 Introduction 181
2 Barriers toward achieving zero waste 182
3 The notion of the "zero waste city" 183
4 The concept of zero waste and the zero-waste city 184
5 Key steps toward zero waste cities: Zero waste management 185
6 Waste treatment and disposal 185
7 Regulatory strategies 186
8 Governance and infrastructure 186
9 Market creation 186
10 Global waste issues and cities: Why zero waste? 187
11 Current and future cities: A crucible of issues—A milieu for innovation and opportunity 192
12 Zero waste: Formation, convergence, circularity, and critique 194
13 Who-how-what-why: Zero Waste? 196
14 Conclusions: Zero waste and the design of future cities 200
References 204
Further reading 212

1 Introduction

Cities are dynamic entities of overlapping social, economic, environmental, and infrastructure systems (UNFPA, 2007). Urban infrastructure serves an increasing proportion of the human population. As cities grow, existing infrastructure is integrated with new approaches and technologies, creating systems that are linked across both physical and temporal scales. Currently, the world's cities generate about 1.3 billion tons of solid waste

Concepts of Advanced Zero Waste Tools
https://doi.org/10.1016/B978-0-12-822183-9.00008-8

per year, and the volume is expected to increase to 2.2 billion tons by 2025. Waste generation rates will more than double over the next 20 years in lower-income countries. However, this current trend of generating waste is not a recent practice; it comes from the very early stages of modern society. So how would it be possible to transform current cities into zero-waste cities? Zero waste means designing and managing products and processes systematically to avoid and eliminate waste, and to recover all resources from the waste stream. Working toward zero waste has become a worldwide movement that motivates changes in design that make it possible to disassemble and recycle products. To put it simply, zero waste means no unnecessary and unwanted waste from a product at any stage of its life cycle. The scope of zero waste comprises many concepts that have been developed for sustainable waste management systems, including avoiding, reducing, reusing, redesigning, regenerating, recycling, repairing, remanufacturing, reselling, and re-distributing waste resources. Hence, a zero-waste strategy is growing in popularity as the best practice. It not only encourages the recycling of products but also aims to restructure their design, production, and distribution to prevent waste emerging in the first place. In recent years, several studies have been conducted to identify the key challenges, drivers, tools, and strategies for achieving zero waste goals.

2 Barriers toward achieving zero waste

Waste management conventionally focuses on products that have reached the end-of-life and thus constitute waste. However, nowadays, the stages occurring upstream in the supply chain are being increasingly considered. Since waste no longer represents the direct interest, this issue is frequently termed resource management.

Waste prevention was assigned the highest priority and should take place before a substance, a material or a product evolves into waste. This aspect aims to avoid or reduce waste in terms of:

- The quantity of waste (e.g. re-use or extension of the life span).
- The adverse impacts of the generated waste on human health following re-use.
- The content of harmful substances in materials and products (i.e. quality of waste).

Reuse is applied to indicate any operation by means of which products or components that are not waste are used again for the same purpose for which they were first conceived. Once again, this action should be undertaken before the formation of waste, thus constituting waste prevention. Directive 98/2008/EC lists preparing for re-use in the second rank indicating that before further reuse can take place a check, cleaning, or repairing operation is commonly required. However, the physical and chemical constitution of the product is not altered.

3 The notion of the "zero waste city"

Cities are over-consuming, and per capita waste generation is relatively higher in high-consuming cities compare to low-consuming cities. Cities attract people because of the economic and social activities and quality of life offered to their inhabitants. However, inadequate urban management, often based on inaccurate perceptions and information, can turn opportunity into disaster (UNFPA, 2007). Currently, many cities are designed and planned based on eco-city concepts, and those cities are designed to deliver a high quality of life to their residents. Completed eco-city projects, such as Vauban Freiberg (Germany), Hammarby Sjöstad (Sweden), and uncompleted projects, for example, Masdar City (UAE), Tianjin Eco-City (China), are designed to offer a good quality of life. All those eco-cities are designed by considering sustainable city design practices. The population density of those completed and uncompleted eco-cities was between 50 and 150 people/ha (Lehmann, 2010a, 2010b, 2010c). However, there can be an argument on the definition of a true "eco-city". For example, a modern city built with a high ecological footprint is not an "eco-city"; moreover, it is not possible to accommodate the entire world's people in the limited global land area in the same design criteria. Many studies have been conducted in different cities to design effective waste management systems aiming at zero waste, including studies in Masdar City (Nader, 2009), Tshwane (Snyman and Vorster, 2010), Taiwan (Young et al., 2010), India (Colon and Fawcett, 2006), Australia (Clay et al., 2006), Greece (Zotos et al., 2008), and England. However, there have been very few studies on a holistic approach to zero waste cities. Cities are not only over-populated and over-consuming, but they also deplete global finite natural resources at a high rate. There is a positive relationship between urbanization and poverty (UN-HABITAT, 2008), and the relationship indicates that sustainably expanding cities is an important factor for global sustainability. How to redesign the existing systems, how to design new products for consumption systems, and how to design new scenarios (Vezzoli and Manzini, 2008) for quality of life are now major questions for planners or researchers.

4 The concept of zero waste and the zero-waste city

Zero waste means designing and managing products and processes systematically to avoid and eliminate waste and to recover all resources from the waste stream (ZWIA, 2004). Working toward zero waste has become a worldwide movement that motivates changes in design that make it possible to disassemble and recycle products. To put it simply, zero waste means no unnecessary and unwanted waste from a product at any stage of its life cycle. The scope of zero waste comprises many concepts that have been developed for sustainable waste management systems, including avoiding, reducing, reusing, redesigning, regenerating, recycling, repairing, remanufacturing, reselling, and re-distributing waste resources. Hence, a zero-waste strategy is growing in popularity as the best practice. It not only encourages the recycling of products but also aims to restructure their design, production, and distribution to prevent waste emerging in the first place (UNECE, 2011).

Most modern societies have been implementing integrated waste management systems to recycle and recover resources from waste. However, the concept of zero waste is not limited to optimum recycling or resource recovery; in addition to that, zero waste requires the elimination of unnecessary waste creation at the first stage of designing a product. Therefore, zero-waste design principles go beyond recycling to focus first on avoidance and reduction of waste by innovative product design and then recycling and composting the rest (City of Austin, 2008).

Fig. 1 shows the key principles of the zero-waste city. With proper implementation of all these principles, current cities could be transformed

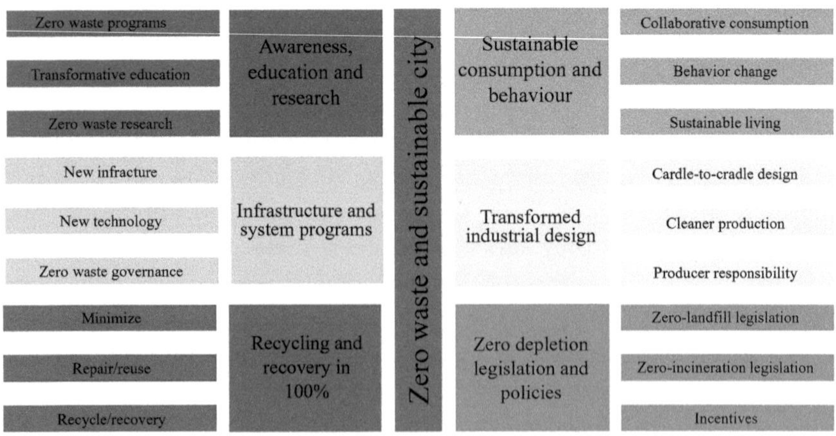

Fig. 1 Drivers for transforming current cities into zero-waste cities.

into zero-waste cities. The key drivers are based on short- and long-term implementation strategies. Awareness and education, behavior change, and systems thinking are long-term strategies, whereas innovative industrial design, legislation, and 100% recycling are the short-term strategies to implement in a city. One of the important aspects of the zero waste city is the conversion of the linear city metabolism to a circular city metabolism.

5 Key steps toward zero waste cities: Zero waste management

The traditional concept of waste as something produced at the end of our consumption activities has to be challenged by the new zero-waste concept. In the zero waste concepts, waste is considered as a transitional phase of resources which can be taken back to the production process or disposed of to landfill. Hence, zero-waste management systems include product design, consumption, and resource recovery Phase*s*. *zero*-waste products, based on cradle-to-cradle design principles, would eventually not produce any waste during their production stages. At the end of the product stage, the life span of the product should be easily extended by repair and thus be ready for reuse again. If not, then the product would easily be dematerialized to reprocess in the creation of a new product. If dematerialization is not possible, then the production process needs to be changed to make it more resources. Finally, the non-usable product would be recovered from the household waste stream and diverted from landfill. The zero-waste management concepts thus integrate both production of products and management of their end-of-life. Hereford, an extended producer and consumer responsibility are very important in zero waste management. Implementing producers' take-back schemes is a resource recovery strategy being applied in many countries.

6 Waste treatment and disposal

Waste treatment and disposal is another key area on the zero waste roadmap. The technology that we use in waste management systems (such as biological and thermal treatment technology) has effective overall waste management systems overall. Incineration of waste may generate heat and energy but the resources that could be potentially recovered are permanently depleted by the mass burn (incineration) systems, and filling waste similarly depletes resources and emits greenhouse gases (GHG) in the atmosphere. Therefore, while thermal treatment and landfill may be temporary waste management solutions, for a permanent and zero waste solution these technologies need to be transformed into zero incineration and no-landfill systems by implementing long-term zero depletion principles.

7 Regulatory strategies

Regulatory strategies are important for sustainable waste management. Regulatory strategies have long been used to encourage or limit certain mass burn technologies. For instance, laws that restricted disposal of organic, combustible, and hazardous waste to landfill imposed higher landfill taxes and provided more incentives would encourage optimum resource recovery.

8 Governance and infrastructure

Experts observe that an effective waste management system reflects good governance. Zero waste governance requires new management approaches and infrastructure. A combination of effective 'hard' and 'sol' infrastructure is required to ensure optimum service to society. Waste collection, storing, sorting and recycling, treatment and disposal facilities are 'hard' infrastructures. Education, regulations, and financial systems are 'sol' infrastructures. Community-based waste recycling centers, hazardous waste collection systems, and separate waste collection for optimum recycling systems are priority areas for governance and infrastructures. A centralized waste data recording system with time series waste data is very important to analyze, assess, and measure the performance of waste management systems. Community recycling centers are landfilled as priority areas for waste recycling of hazardous, non-hazardous, and bulk waste.

9 Market creation

Our current globalized economy is predominantly driven by market systems. Since waste is part of everyday life, market-driven waste management solutions are vital to improve current waste management systems. In a positive market situation, waste management is easier and more cost-effective than in a negative market situation. In most of the developed countries, due to high labor and recovery costs, waste is sent to developing countries by accepting the negative consequences of resource and environmental depletion.

10 Global waste issues and cities: Why zero waste?

Resources from all over the biosphere are described as being, "funneled" into the world's cities to meet the rising consumption, driven by accelerating globalization, urbanization, and affluence (Krzeminska et al., 2017; Girardet, 2000). A commonly cited metric illustrating the associated imbalance and exploitation is that cities occupy just 2% of the earth's land surface, yet use over 75 percent of its resources and discharge corresponding proportions of waste (Girardet, 2000). Similarly, urban communities currently account for over 70% of global energy-related CO2 emissions (Edenhofer et al., 2014). Extrapolating this exploitation discharge correlated further, cities appear as the apex of the "anthroposphere", which Manahan articulates as centrally embedded in a rubric of material, energy and waste exchange within the dynamic interaction of atmosphere, biosphere hydrosphere, and geosphere (Manahan, 1999).

Such reporting draws a stark distinction between the extractive, carbon-intensive, lineal, disposal-orientated human systems, and the "biological analogy" (Ayres and Ayres, 2002) and "ecosystem" design metaphor (Isenmann, 2008; Korhonen, 2004) offered in, what are popularly interpreted as the infinitely sustainable, solar-powered, zero emissions, and circular metabolism of natural systems (Lehmann, 2011a, 2011b; Lehmann, 2010a, 2010b, 2010c). The functionality of urban systems is said to determine whether the waste outputs of cities, discharge to atmosphere, or gets deposited in dumps/landfills, or accidentally/deliberately litters the landscape, before accumulating in rivers and the ocean (Hoornweg et al., 2012). In respect of more sustainably managing the globally significant urban resource/waste stocks, flows, and sinks, the development of sustainable future cities is cited as requiring a regenerative design mindset and fostering new technical and organizational solutions, which bio-mimic nature's inherently successful—circular design (Benyus, 1997; Garcia-Serna et al., 2007).

Striving for zero-carbon transport, energy, and building systems (i.e., "nZEB") (Riffat et al., 2016), resource conservation and efficiency, and the beneficial recycling and reabsorbing non-toxic non-polluting water (Verstraete and Vlaeminck, 2011) and waste flows, as a resource (hence enhancing the quality of life and the long-term environmental sustainability of the whole system) are identified as critical challenges within future city discourse (Verstraete and Vlaeminck, 2011; Porse, 2013). Zero waste (inclusive of future zero-waste city models) has been described as a pathway being "forged" toward a desirable long-term goal (Bartl, 2011) and in some sectors almost achieved (Mezher, 2011). However, zero waste is also regarded negatively by some, as a potentially harmful myth (Premalatha et al., 2013). Notwithstanding this spectrum of reporting, a range of integrated green urban design principles, such as sustainable design and circular urban metabolisms, have been discussed as being central to realizing the concept of a zero-waste city (Hoornweg et al., 2012). Given the conflicting assertions that the phenomenon of zero waste affords a critical opportunity to address waste issues (Hussain, 2020), a starting point in exploring these claims, is to examine current global "progress to date" in managing waste.

A cluster of international reports describe the issues that are associated with waste, as of becoming a globalized public and environmental health emergency, necessitating an urgent, internationally coordinated, comprehensive, and effective response (Mavropoulos et al., 2015, 2017). The environmental and social consequence of humanities failure to effectively manage waste has resulted in some of the most polluted and poverty-stricken places on Earth (Mavropoulos et al., 2017). Whilst this syndrome is often localized and most concentrated around (mega) cities (UN-Habitat, 2010; Abarca Guerrero et al., 2012), the interrelated aquatic and atmospheric dimensions of impacts of terrestrially generated waste are now being registered across the entire global biosphere (Ryan et al., 2009; Hodzic et al., 2012).

The World Bank reported that the 2012 baseline of 1.3 billion tons of municipal solid waste (MSW) generated by cities globally, is projected to double by 2025 to 2.2 billion tons pa (Hoornweg et al., 2012). The current trajectory of growing population, urbanization, and consumer demand underwrite such projections (Mavropoulos, 2010a, 2010b; Mavropoulos, 2014; Troschinetz and Mihelcic, 2009). Given this, it seems unlikely that the critical challenge of reducing waste generation (i.e., located as the top priority of the "5R" waste hierarchy (i.e. firstly: reduce, reuse, recycle, recover energy, and then lastly residual disposal)) is, under "business as usual" conditions, immediately achievable (Hussain, 2019). Concerningly, it has been reported that, unless aggressive sustainability scenarios are successfully implemented, "global peak waste" might not occur until 2100 (Hoornweg et al., 2014). Increasingly, the interrelated dimensions of the waste issue, (i.e., ocean plastics, disaster waste management, chemical toxicity and dissipation, food-waste, organized crime, nuclear waste, and emerging "NBRIC" (i.e. nanotechnologies, biotechnologies, information and communication technologies (i.e. WEEE), robotics and cognitive sciences (Graedel and Allenby, 2010) is attracting media reporting and a correlated escalation in public awareness and alarm. This breadth and diversity of waste issues is overlain by systemic causalities, such as history, geography, infrastructure and technology, vested interests and ideology (i.e., privatization (Iskandar and Tjell, 2009)), and individual and collective cultural and socio-economic imperatives (Wilson, 2007; Marshall and Farahbakhsh, 2013; Gutherlet, 2010; Pongrácz and Pohjola, 2004; ISWA, 2015), which adds to associations of "super wicked" (Krausz, 2012; Levin et al., 2009; Levin et al., 2012) complexity and intransigence.

In terms of the global provision of "residual disposal" (i.e., the supposed least priority, at the bottom of the 5R waste hierarchy) the efficacy and outcome to date of the conventional waste management paradigm and practice, also raises questions. The International Solid Waste Association's (ISWA)—"Global Waste Management Outlook" (GWMO) aligns with other similar reporting, in estimating that, between 2 and 3 billion people live below the most basic waste management system benchmarks of collection and controlled disposal (D-Waste, 2013; Wilson et al., 2015a, 2015b). Aggravating concerns around the pollution and climate change impacts of systemic failings in global waste management, reporting indicates that the default disposal "treatment" for approximately 41% of global waste is uncontrolled burning (Wiedinmyer et al., 2014; Thompson, 2014).

The critically important global ISWA programme seeking to rectify this syndrome (Mavropoulos et al., 2017) has set challenging goals (i.e., "As an initial step, aim to: achieve 100% collection coverage in all cities with a population more than 1 million, eliminate open burning of municipal solid wastes and similar wastes, and close large open dumps, replacing them with controlled disposal facilities" (Wilson et al., 2015a, 2015b)). Achieving these goals represents a key initial benchmark in modern "integrated solid waste management" (Wilson et al., 2009; ISWA, 2015). However, it is important to recognize that achieving those baselines, is just the starting point for the envisioned transition to holistic, sustainable resource conservation, and material circularity, which advocated in, for example, "circular integrated waste management systems" (CIWMS), zero waste and a circular economy discourse (Zaman and Lehmann, 2011a, 2011b; Ellen MacArthur Foundation, 2013, Murray, 2002; Silva et al., 2017; Cobo et al., 2018).

Whilst it can be accepted that "the world can't recycle its way out of waste" (Mace and Szaky, 2016). Equally, the common scientific rhetoric offered by the USEPA (2013) underwrites the growing ubiquity and popularity of recycling today. Keynote environmental commentators similarly link the benefit of recycling to the challenge of addressing climate change. Stern argues that, because recycling makes such major and under-appreciated contributions to reducing GHG emissions it is one of the "best kept secrets in energy and climate change" policy (Stern, 2009). Such glowing appraisals has been more recently "reality checked" by China's successive "Green Fence", "National Sword", and "Blue Sky" import policies applying to recycled materials, which has sent shockwaves through global recycling markets (WasteMINZ and Eunomia Research and Consulting, 2018). Overall, the importance and positive opportunity of "recycling citizenship" demonstrated by the informal sector and communities across the global spectrum of socio-economic development, is now well established (Silva et al., 2017; Gutherlet, 2008; Wilson et al., 2009; Wilson et al., 2006; Velis et al., 2012; Allen et al., 2012.

However, in spite of the significant environmental and social opportunities that are attributed to recycling, it is estimated that globally, currently only one-quarter to a third of the total 3.4–4 billion tons of MSW and industrial waste produced annually, is recycled (Chalmin and Gaillochet, 2009). So in summary, international waste data indicates that, after over four decades of significantly investing in the widely accepted principles of the "waste hierarchy", there are still significant barriers in realizing the stated: top (reduce), middle (recycle), or even lowest (residual disposal) priorities. Whilst conventional waste management theory, distilled into the near-universal rubric of the waste hierarchy, clarifies our priorities and can be seen as having catalyzed a measure of progress, overall we are yet to globally actualize this principle and appear to be "entangled/trapped" in limitations of this paradigm (Pietzsch et al., 2017; Bartl, 2014a, 2014b, 2014c; Van Ewijk and Stegemann, 2016; Pollans, 2017).

The net result is that most of the resources which flow through the global economy still transit via the destructive and polluting linear model, variously described as—"take-make-waste" (Jessen, 2003) / "dispose" (Pietzsch et al., 2017; Ellen MacArthur Foundation, 2013). Evidencing this, socio-metabolic research, which assessed the degree of circularity of materials flowing through the global economy, describes this as currently, only in the early stages (Ghisellini et al., 2016; Haas et al., 2015). Currently, the development of a more "circular economy" is limited by rapid growth in "socio-economic stocks", a focus on recycling rather than reuse/reduction and an estimated 44% of processed materials that are incinerated to "provide energy" (Pietzsch et al., 2017; Haas et al., 2015), and hence, exit rather realize economic circularity.

The zero-waste movement (Anderson, 2011) can be viewed as one of a cluster of sustainability actors, which both highlight and respond to the nexus of failure, inertia, and growing sense of crisis, that is associated with the conventional waste management paradigm (Silva et al., 2017; Hannon, 2015). The zero-waste movement encompasses a range of perspectives and approaches ((Pietzsch et al., 2017) and can be regarded as a neologism, residing in a busy "eco-ideas marketplace", alongside interrelated and complementary theses on how sustainable development can be engineered (Pietzsch et al., 2017; Haas et al., 2015; Glavic and Lukman, 2007). For example, whilst disciplines, such as industrial ecology (IE), urban metabolism (UM), and bioeconomy (BE), and the movements for a "circular economy" (CE) and zero_waste each arise out of differing: perspectives, personalities, and intellectual traditions, the appearance of shared cognitive DNA seems clear (Veleva et al., 2016; Hannon et al., 2018). These movements are conceptually aligned and complementary in seeking to confront and re-design and replace the current "exploitative", lineal economic model with progressively more cyclical and sustainable resource management, where anthropogenic systems "bio-mimic" the modeling of natural systems (Benyus, 1997; Murray, 2002; Hawken et al., 1999).

However, in this sphere, zero waste also has a unique identity and assumes a distinctive role, articulated in the broadly accepted, peer-reviewed definition offered by the Zero Waste International Alliance (ZWIA) (ZWIA., 2009). In the adoption of provocative terminology, a campaign posture and in advocating for a hyper-aspirational continuum of innovation, zero waste seeks to confront the perceptions of normalcy and intractability around waste (Hussain and Mishra, 2019). The embrace of dissent and activism in the framing of zero waste, alongside the embrace of community/NGO involvement and the economically redistributive aspects, is why the movement is simultaneously controversial, and arguably indispensable (Pollans, 2017; Lombardi and Bailey, 2015).

Encompassed in the prickly opposition to incineration and landfill, zero waste seeks to refute and disrupt the prevailing normalization of waste and our "throw-away society, as a relatively recent socio-economic construction, which can and must, be redesigned (Waste Watch, UK, 2004; Herbert, 1998). Zero waste directly confronts the waste management industry's twin bury and burn profit centers, on the basis that perpetuating our "flame, flush or fling" (Seadon, 2010) disposal mentality, ultimately binds human society to linear material flows, rather than enables the development of a more circular economy. Rather than extolling the supposed technical progress of reforming disposal systems (such as sanitizing, or optimizing landfill or extracting energy from incineration) zero waste regards these "developments" as confirmation of societal capture to a failing and unsustainable socio-economic model (Seadon, 2010; Connett, 2013). The thwarted global progress toward genuinely sustainable material resource management is the central provocation catalyzing the global search for alternative modes for generating innovation and progress. Within this spectrum of activity, a growing cohort of organizations and practitioners choose to self-identify, under the heterogeneous brand of zero waste.

11 Current and future cities: A crucible of issues—A milieu for innovation and opportunity

Historically, the creativity, social connectivity, and productivity that are facilitated by cities have been central to human advancement and technological development. However, this progress is also associated with negative environmental consequences, including waste-related issues (Riffat et al., 2016). The confluence of globalization, megacities, and waste management provides a lens into anthropogenic dysfunction and crisis (Mavropoulos, 2014). It appears likely future cites (including and especially African and Asian megacities (Mavropoulos, 2014; Cohen, 2006; Oyoo et al., 2011; Swapan et al., 2017) will be a critical focal point for addressing the challenges associated with population growth, globalization, climate change, and sustainable development. Conversely, cities will remain the prime locus, where negative consequences manifest and further undermine environmental sustainability, social and economic stability, and well-being (Oyoo et al., 2011). Illustrating the problem-solution dichotomy of future cities, the triangulation of "city, people, planet" is also cited as providing a hotbed of waste-related liveable city innovation (Girardet, 2000).

As the culmination of multiple system vulnerabilities (i.e., acute poverty, accentuated by extreme social inequality, escalating urbanization, socio-economic development latent proximity to geographical hazard zones and regional environmental degradation, appended to the consumption of material resources, energy, and water) (mega)cities form apices of global risk (Cook and Swyngedouw, 2012). However, despite the exceeding ecological footprint and concentration of systemic social issues, it is wrong to assume that cities are automatically bad for the environment. In contrast, future (mega) cities should be understood as a crucible of opportunity and "political ecology" necessary for research-led hyper-innovation and the enabling social cohesion, which might underwrite necessary reform and progress (Mavropoulos, 2014; Cook and Swyngedouw, 2012; Earth Science for Society, 2005). For example, there appears to be a positive correlation between addressing waste alongside other critical "future city" issues, such as sustainable food systems. Illustrating this, urban agriculture is posited as providing an opportunity to reduce the outright generation of food waste (and the associated expending embodied energy), as well as by being the receiving environment (hence enabling localized nutrient cycling) for the compost and other products of organic recycling systems (Ackerman et al., 2014).

Cities provide, almost irrespective of geography, location, culture, and history, etc., a characteristic "template" of environmental and social problems (Krzeminska et al., 2017). As such, cities become vectors for the cognition and design rhetoric of aspiration and solution-seeking (aka "eco/solar/smart-city", "green-urbanism", "bionic", or similarly with water as a focus, "spaceship") (Krzeminska et al., 2017; Fairburn et al., 2014; Lehmann, 2010a, 2010b, 2010c). Within such futuristic and envisioning design paradigms, future cities are seen offering the disruptive new formations of social-political ecology and as providing living laboratories of, for example innovative and transformative governance/management, smart technologies/infrastructure (i.e., ICT enabled internet of things inc. in respect of waste (Batty et al., 2012) to foster zero-energetic, zero-waste, environmentally sustainable self-sufficient food security, and symbiotic industrial ecologies (Bach et al., 2010; Naphade et al., 2011).

In respect of the design imperative for future cities, zero waste can be viewed as a "creative milieu", functioning across the spectrum between, the ZWIA's formal definition and a "wild-west" of interpretive miscellany. Zero waste exists in a tension between the ZWIA's genuine attempts to quality assure and preserve the integrity of the concept and the creative freedom that is required to drive the quantum innovation that is needed to address the escalating spectrum of waste issues, which demand a continuum of locally appropriate responses (Wilson et al., 2015a, 2015b; UNEP et al., 2013). Leveraging off the mutability and envisioning function of zero—as a "stylistic" for innovation, zero waste can be seen as an optimistic meme, for a "future and solutions focused" freedom of thinking (Hannon, 2015).

For example, spheres of acute innovation with potentially far consequence, which are colonizing, re-interpreting and are simultaneously manipulating, stretching and actualizing the plausibility of zero waste are nanotechnology (Tsuzuki, 2010; Nanotechnology Development Blog, 2013; Gibbs, 2011) 3D printing (Campbell et al., 2011; Chonga et al.,

2015; Taylor, 2014) and in the context of space travel (ESA, 2013; Morgan, 2013), lothing (Electrolux, 2014; Korge, 2014), housing (Wainwright, 2014). Zero waste can also be interpreted as being part of a sixth wave of innovation in waste management systems and continues to be debated, contested, and to evolve across its globalized contexts and interpretations (Electrolux, 2014; Korge, 2014).

Described historically as a "dangerous idea", the paradox of zero is cited as inspiring the disruptive "zeronautics" that are capable of breaking the sustainability barrier (Seife, 2000; Elkington, 2012). The idiosyncratic coinciding of the terms waste and zero signifies a direct challenge and the deliberate confrontation of both, the normative function and implication of the non-neutral language of waste and the deeply vested industrial triumvirate (aka "Take-Make-Waste" (Oyoo et al., 2011), dominating the core of the global current economic system. The discourse and community of zero waste in the respective theoretical nurture and actual participation in waste activism, denotes a rejection and stark disassociation from the entrenched role of traditional waste management, as an enabler of the economic primacy given to consumption and capitalist growth models (Mavropoulos, 2014; Cohen, 2006; Oyoo et al., 2011). Even before encountering any of the numerous actual practical barriers to influencing "modes of governance", the word couple—zero and waste, directly contests the embedded connotation of waste, relative to the dominant political rhetoric and reality (Murphy and Princetl, 2013).

12 Zero waste: Formation, convergence, circularity, and critique

Zero waste is also described as a unifying concept that embraces the diversity of measures, experiences, and interpretations arising in industrial, municipal, activist/community, development, and policy/government spheres of practice (Oyoo et al., 2011). It can be argued that the truest holistic comprehension of all that zero waste is perceived to be, is expressed cumulatively across all the various respective media (Inc. social) and other avenues of publication, including, but not exclusively, academic literature. However, in the latter respect, there is now a cluster of substantive review articles, providing an improving summation of the evolving and overlapping descriptions of the continuing flux in zero waste development (Stern, 2009). It can also be recognized that deeper analysis, further clarity, and case study research (Stern, 2009; Campbell-White, 2016) is required. For example, specifically around policy implications, quantifiable success factors and the relational distinction to "zero waste to landfill" (ZWtL) and waste to energy (W2E) (Campbell-White, 2016; Gharfalkar et al., 2015; Themelis, n.d.; Krausz et al., 2013).

Given the innate challenge, complexity, and formative inconsistencies of this sphere, alongside recognition of the role that "contested and negotiated narrative" plays in policy formation, to what extent are critiques, such as that zero waste is populist, over-simplified, reactionary, and/or too extreme (Campanelli, 2011; Ragossnig, 2006; Bartl, 2014a, 2014b, 2014c; Themelis, 2009) justified? Interestingly, there is now nothing too unique or special about the ideal of zero waste. Illustrating this, the mainstream International Solid Waste Association's (ISWA) vision (See: ISWA, 2015) is to work toward an "earth where no waste exists" (Oyoo et al., 2011). Similarly, the Solid Waste Association of North America (SWANA) both acknowledges "the paradigm shift towards of zero waste." and offers training in support of zero waste as "... real trend" (Trotti, 2012). The world's largest waste management company is cited as framing zero-waste service provision within an integrative, synergistic future perspective (Humes, 2012). So, whilst many seem to adopt and agree with a generalized vision and some of the rhetoric of zero waste, there is a spectrum of interpretations and debate around what represents, "faux" (Palmer, 2013) versus "authentic" (Zero Waste Canada, n.d.) zero waste, in what is a globalized "free-market" of ideas and activity.

It can be argued that "business as usual", token, or even outwardly positive incremental change to the deeply embedded causalities and the manifest issue of waste, perpetuates the growing risk of stumbling over a tipping point into irreversible pollution and climate change impacts (Lenton and Ciscar, 2013; Schneider, 2009; Hansen, 2009). In contrast to the risks that are associated with the status quo, effectively tackling, even just the most acute baseline waste issues, has been associated with the potential, across the global economy, to reduce "GHG emissions by 15–20%" (This is only one of many benefits for example: Aside from the significant opportunities to reduce toxicity, pollution, the "estimate of worldwide potential for new jobs in the circular economy is 9 to 25 million. Prevention of the 1.3 billion tons of food waste generated per annum enough to feed all the undernourished people in the world twice over", whilst addressing "more than half of the high-level sustainable development goals within the Post-2015 Development Agenda" (Mavropoulos, 2014).

Whilst, accepting that the global problem of waste is an extreme challenge, it is also important to recognize that the cost to society of not addressing the most acute issue of dumpsites is cited as exceeding the "financial cost per capita of proper waste management by a factor of 5–10" (Mavropoulos, 2014). Relative to the wasteful status quo, generating progress presents as a sound economic investment (Inc, 2009; Hood, 2013). The apparent strength of the socio-economic and environmental opportunity of addressing waste issues highlights an important question. Why positive progress has been so difficult to catalyze in this sphere? Key international thinkers now reconceptualize waste as a symptom and physical artifact of a fundamental and unresolved failure in socio-economic design (Ghisellini et al., 2016; Haas et al., 2015). The roots of this design failure originate in the post–WWII reconstructionist period, where faulty perceptions of infinite resources, consumerism, and the "throw-away society" were socially engineered into the DNA of what has become the dominant political and economic ideology (Porritt, 2007).

Today, the movements for zero waste and a circular economy advocate for a transition away from this wasteful and polluting socio-economic model, premised on lineal resource flows, environmental exploitation, and excessive disposal. A "circular economy" has been articulated as being: "regenerative by intention and design (Hussain and Kharisov, 2017) eliminates the use of toxic chemicals. aims for the elimination of waste through the superior design of materials, products, systems, and, within this, business models" (Haas et al., 2015). Advocates of a circular economy note that: "nature operates according to a system of nutrients and metabolism in which there is no such thing as waste" (Ghisellini et al., 2016).

Similarly, the discipline of industrial ecology, like zero waste, also seeks to emulate the "ecosystem metaphor", which recognizes that "nature is a zero-waste system. Nature recycles everything" (Snow and Dickinson, 2001). In rejecting the "concept of waste" (Krausz, 2012) and seeking to loop the "technosphere back on to itself" (Bourg and Erkman, 2003) industrial ecology can be seen as combining a bio-mimicry of natural systems (Levin et al., 2009) and the syntax of recycling, in progressing the "ultimate industrial ecology goal of zero waste" (WMAA and AIEN, 2013). Similarly, the discourse and practice attributed to the global bioeconomy movement shares in and illustrates the ubiquity, ideal, and rhetoric (i.e., "green", "cycles", "zero", "nature", etc) of this sustainability construct (Loiseau et al., 2016; Maciejczak, 2015; El-Chichakli, 2016; Pfau et al., 2014).

13 Who-how-what-why: Zero Waste?

It has been argued that solid waste management in the world's cities is not only "the most important municipal service", but because of issues, such as relative cost, projected increases in generation rates, and the interrelationship with other local considerations (i.e., transport, informal sector, flooding, pollution, and public health), effective waste management in cities is viewed as a key prerequisite enabling other sustainable urban development activity (Hoornweg and Thomas, 1999). This and other insights into the city—waste issue nexus, support the view that globally, waste is amongst the most challenging and complex anthropogenic problems confronting communities today. Zero waste case studies provide a degree of both, sagacity and future optimism around the challenges of solid waste management in the world's cities, in particular when community involvement and responsibility is recognized and enabled in the quest for local solutions and progress (Asian Development Bank, 2017). So, amongst the spectrum of theories, imperatives, and investment directed at waste issues globally, how significant and effective is the zero waste contribution?

From the name down, the zero waste movement deliberately embodies confrontational terminology and appears to embrace the risk of confusing infinite inspiration with enigma and antagonism. Given the depth of international consultation, which underwrites the terminology selected for the ZWIA definition, it seems that the movement strategically accepts controversy and embraces the attendant criticism and push-back. Whilst, some of the criticism that was directed at zero waste appears overwrought and misguided, a review of the critique of zero waste illuminates debate, which informs and challenges both conventional waste and zero waste schools of thought. Despite the appearance of being "black n white", the zero waste movement appears to catalyze a spectrum of new activity, experience, and learning, which spans "revolutionary" to "evolutionary" worldviews (Seife, 2000; Elkington, 2012; Bartl, 2014a, 2014b, 2014c; Amengual, 2014; Dobbs et al., 2013).

Whilst the critical importance of the 3Rs is widely accepted and is supported in the practice of zero waste, conceptually zero waste seeks to be reframed beyond the constraints of conventional waste management perspectives (Ryan et al., 2009; Hodzic et al., 2012; Anthony and Liss, n.d.). This aspiration is expressed in the ideal of seeking to "eliminate", rather than forever just managing waste, as discarded material flow preordained for disposal (Snow and Dickinson, 2001; Chandavarkar and UNDESA/UNEP, 2010). Zero waste seeks to reimagine the default setting by which waste is inevitably characterized a "problem" (i.e., something to be quickly gotten rid of) into being reconceptualized, as a "resource" framed opportunity to be exploited (Pietzsch et al., 2017; Haas et al., 2015).

Zero waste directly challenges the assumption of waste as a valueless and an unavoidable byproduct of consumption. In contrast, zero-waste considers the whole of product lifecycle and whole of economic system viewpoint. Accordingly, commentators describe zero waste as a "manifesto for the redesign of the material economy" (Glavic and Lukman, 2007), which seeks to culminate in a "2nd green industrial revolution" (Murray, 1999; Williams, 2014). Speaking from the perspective of an industrial economist, Murray's challenge is that waste has become an issue, which is "too important to be left to the waste industry" (Murray, 2002). The author compellingly interweaves the idiom of waste, into the mainstream conversations of climate change and resource management in observing that, "as a pollutant waste demands control. As an embodiment of accumulated energy and materials, it invites an alternative" (Murray, 2002).

Derivative of the way the word waste, simultaneously functions as a noun, verb, and adjective, zero waste is used and interpreted differentially, according to worldview, context, and objective. Interpreted inclusively, the zero waste movement is a heterogeneous global community of practice, demonstrating many of the tensions and contradictions incumbent to an emergent, holistic, and heavily contested worldview (Stern, 2009). An overview of zero waste literature shows zero waste arising out of industrial, municipal, and activist/community experiences with apparent precedence and emphasis in that order (Hannon, 2015; Hestin et al., 2010). Each distinct worldview contributes to and shapes the narrative, as well as a sense of juxtaposition attributable to zero waste (Hussain and Keçili, 2019). Whilst it can be recognized that the origins of zero waste success were pioneered in an industry setting, (Hannon, 2015) today endorsing and empowering the informal sector and grass roots-community based initiatives, is seen as instrumental to the future of zero waste (Van Vliet, 2014a, 2014b, 2014c; Chaudhary, 2013). Adding to this diversity and debate, zero waste policies and programs are being developed across the spectrum from developing to developed socio-economic settings (Hill et al., 2006).

At its most deconstructed, zero waste strategy can be understood as involving phased and synergetic "upstream" (designing for prevention-avoidance) and "downstream" (confronting and reforming waste policy and industry practice) conceptions. The upstream sphere of zero waste is where the issues around current products, production, consumption, and urban systems are sought to be addressed through transformative design and social innovation (Lehmann and Crocker, 2012; Levitzke, 2012). Allied to this, the downstream expression of zero waste, is where conventional waste management's theory, policies, and practices are contested and sought to be radically reimagined and reformed. In practice, the implementation of zero waste in a holistic community—municipal setting, has also been cast in a variety of frameworks, for example, as a "10 step" and as a 3 phase, 21 action (pre-conditional on six core infrastructure supports) community "roadmap". Zero waste has also been described as contingent on the integration of five key principles for transforming existing, wasteful, and polluting systems of production, consumption, and disposal (Zaman and Lehmann, 2011a, 2011b).

Alongside extending and formalizing producer responsibility and revising legislation to support recovery and recycling vs disposal (including for waste material flow data) (Murphy and Princetl, 2013), the transition to a cradle-to-cradle—zero-waste economic model, also requires a transformation in our concept of citizenship, globally, to include sustainable and responsible patterns of purchase and consumption. In this modality, traditional "end of pipe" waste management will be reshaped by the new environmental services demanded by closed-loop recovery and recycling systems. A core function of future (zero) waste management systems, is to cycle resources between producers and consumers, in a resource-efficient and circular zero-waste economy (Morgan, 2013). In this framing, it can be argued that zero waste is less of a competitor to traditional solid waste theory than it is a synergist, catalyzing a shift up into the top, "largely uncharted" priorities of the waste hierarchy.

The absolutist misnomer, promoted in some of the criticism of zero waste, is perhaps the easiest misinterpretation to clarify. Put simply, zero waste is less about absolute zero, than it is about generating focus and urgency for progress. This is evidenced in the following quotes, by recognized zero waste commentator: "It's important not to get hung up on the zero. No system is 100% efficient" and "We have got to get over this idea that we are talking about 100% diversion" (Leroux, 2001). "Beyond Recycling! Zero Waste ... Or Darn Near" (Lombardi, 2001). "Zero waste is not a literal target. It may not be possible to eliminate every item of waste stream—but we will not know how far we can get unless we try. If do not strive for zero waste, we will continue to make only incremental progress to stem this tide of waste" (Dimino and Warren, 2004). "Zero waste is a goal that we know we can't really get to. But it also is a process with very clear environmental, social and economic benefits. By working through the process you get closer to the goal" (Seldman cited in (Leroux, 2001)).

Nor do zero waste advocates seek to redraft the second or fourth laws of thermodynamics, according to their own flight of impossible/unscientific fancy (Clift, 2004). Put simply, it can be assumed that nature does not defy her own thermodynamic Laws. Zero waste's injunction is not to exceed or de-bunk thermodynamic principles, it simply aligns with the now ubiquitous IE/UM/BE/CE rhetoric, which society should rectify failing economic principles and practices to conform to the sustained success and circularity modeled by nature.

14 Conclusions: Zero waste and the design of future cities

Cities are typically, concentrated "hot-spots" of production and consumption, which results in an unsustainable ecological footprint (i.e. greatly in excess of bio-capsity), concentric environmental impacts (i.e., altered land-use and biochemical cycles) and embedded dependencies on other cities and regional hinterlands for the supply of materials, energy and for the disposal of waste (Agudelo-Vera et al., 2012a, 2012b; Grimm et al., 2008). However, cities are being re-conceptualized as a critical locus of future resilience and sustainability, through, for example, the potential of "urban harvesting"/"above ground mining" to recirculate the globally significant reservoirs/sinks and flows of secondary resources embodied in the dynamic of cites (Agudelo-Vera et al., 2012a, 2012b; Graedel, 2010; Greenfield and Graedel, 2013).

The urban design ethos of future zero-waste cities extends beyond just the technical engineering of renewable energy and closed-loop circular material eco-systems (aka the tripartite aims of "zero waste/fossil fuel use/emissions") to also encompass holistic environmental, and social sustainability and instrumental principles for healthy communities (Lehmann, 2010a, 2010b, 2010c). Designing for sustainability, permeates all 15 interrelated principles of "green urbanism" (Lehmann, 2010a, 2010b, 2010c) and seeks to embed change and ongoing evolution in every aspect of designing, building, operating, maintaining, and renewing and recycling in future zero-waste cities. The holistic design process for future zero-waste cities, prioritizes social transformation focused on consumption (to in the first instance avoid creating waste) and reconceptualizes (and accordingly seeks to enable the analysis and management of) "waste streams" as a valuable flow of material resources (Lehmann, 2010a, 2010b, 2010c; Lehmann, 2011a, 2011b). Strong industry and political leadership, new regulatory and policy frameworks, affordable technologies, infrastructure and programs, and effective education/awareness-raising and R&D are cited amongst the requirements for the zero waste city concept to be realized, rather than subverted as shallow, "technical quick fix, idealistic utopianism" (Lehmann, 2010a, 2010b, 2010c; Lehmann, 2011a, 2011b; Zero Waste Europe, 2017).

As attention shifts from just focusing on reducing supply-side industry emissions, toward understanding the equally significant and commercially attractive demand-side emission reduction opportunities (i.e., "as much as 296 million tons CO2 per year in the EU by 2050, out of 530 in total—and some 3.6 billion tons per year globally. Demand-side measures thus can take us more than halfway to net-zero emissions from EU industry" (Enkvist and Klevnas, 2018)) inherent to a more circular economy (i.e., facilitating better use and reuse of existing material resources), the financial benefits are becoming more clearly recognized and understood (Enkvist and Klevnas, 2018).

Accordingly, as the meta societal shift from traditional, linear, sanitary treatment/disposal based "Integrated Waste Management Systems" (IWMSs), toward "circular IWMSs" (aka CIWMSs i.e., combining waste and materials management) and then further still, into authentic, holistic future zero-waste cities, and circular economies, the "business case" for the social and regulatory interventions to drive this transition will continue to emerge and strengthen. The zero-waste movement, whether reflected in the upstream aspiration of the zero waste design community (i.e., seeking to transform future materials and products which become discards), or the reformative downstream activity of zero waste practitioners (seeking to circularize and upcycle discarded material flows), appears as an integral to, and as an instrumental driver of future sustainability (Crocker and Lehmann, 2012).

Possibly the best argument for the confrontation, innovation, learning, and progress that the zero waste movement contributes to the long-term global challenge of sustainable resource management is non-technical. Inherent to the naturalistic ecosystem metaphor, which inspires so much of the "design think" in this sphere, are the interrelated concepts of bio-diversity and system resilience. A fundamental argument in favor of what the zero waste movement adds to the broad community of effort seeking to resolve waste issues is conveyed in statements such as: "there is no right or wrong answer" or "no one size fits all". Consistent with this advocacy, the zero waste movement can be interpreted as, enhancing the "bio-diversity" of ideation focussed on waste and as, expanding reach and resilience of innovation-seeking, to address waste issues.

It can be argued that, in a post normal era of science (PNS), "uncertainty, value loading, and a plurality of legitimate perspectives" should be accepted and normalized, rather than rejected, stigmatized, and marginalized, as can occur when traditional disciplinary worldviews dominate science practice (Funtowicz and Ravetz, 2003). The zero-waste movement demonstrates the PNS outsider characteristics of: "learning by doing and doing by learning", a sense of transgression and re-assemblage, a multi-actor heuristic, and a lack of fixed typology. In these and the associated practical contradictions and cognitive tensions around transcending, "futuring", and "continuum" above normative short-term, tactical obsequiousness to disposal, zero waste can be recognized as fitting the PNS descriptor, of being a post-normal sustainable technology (Frame and Brown, 2008).

Contemporary descriptions of, and guidance for, future zero waste practices evidences both, "pick and mix adopting and adapting solutions" and endorses local creativity, community engagement, and critical thinking in the outworking core principles. The heterogeneous and evolving global spectrum of industry and community based zero-waste practice reflects local diversity and expands the reach and resilience of the "ecosystem" of responses, seeking to address the human-derived waste crisis. Given the previously discussed, limited "progress to date" reported in global waste data, it appears humanity is yet to categorically resolve the challenge of waste. Until we do—more open-mindedness and other avenues of exploration and innovation, appear essential. In adding grist and diversity to the necessary spectrum of debate and practical initiatives, exploring future change, zero waste is part of the more and the other.

As a grass-roots, countercultural movement, for the most part, zero waste operates on the economic periphery, outside of the dominant "modes of governing" (still firmly adhered to the "disposal paradigm") and in the face of intense competition from vested interests with pejorative control over material flows and flux of ideas in media and policy (Stern, 2009; Campbell-White, 2016). This makes the emerging achievements and experience of zero waste practitioners, all the more interesting and potentially valuable, as we do not yet know what individual, or combination initiatives might catalyze the requisite breakthroughs and wholesale environmental progress. The New Zealand zero waste story suggests, treating with caution, the voices seeking to dissuade and cull the "biodiversity" in our mix of approaches to community and industry engagement in environmental sustainability.

Emerging case studies of zero waste innovation and good practice highlight that, this is potentially a sphere of environmental action, where significant progress can be generated (Campbell-White, 2016; Klein, 2014;

ZeroWIN, n.d.). It can be argued that the zero waste movement's growing track record (i.e., promoting and progressing toward stated goals, in relatively short timeframes (Lombardi, 2001), makes this an important sphere of activity (Lombardi, 2001) for mitigating climate change, implementing UN's SDGs and progressing toward a more circular economy (Stern, 2009; ISWA, 2009). New Zealand's polarized experience around zero waste, underscores the importance and purpose of this article, which is to, further explicate the interdisciplinary phenomenon of zero waste. In illustrating formative examples of waste/zero waste focused "living labs" approaches, seeking to cogenerate innovation in support of the concept planning and practice of future zero waste communities/cities, the authors reiterate prior calls for further elaborated case study analysis and interdisciplinary research and development in this sphere (Hoornweg et al., 2012).

From this discussion, it is clear that the zero waste index provides a better picture of the overall waste management performance of a city than the diversion rate. Moreover, a 100% diversion of waste from landfill would be a milestone for a waste authority but would not necessarily achieve zero waste goals. The diversion rate does not give an indication of resources that have been recovered and substituted, which eventually avoids the extraction of further resources. The zero-waste index forecasts the amount of resources that are recovered from the waste streams and substituted for virgin materials. In addition, the ZWI also forecasts the demand substitution of energy, water, and emissions by the waste management systems.

The overall performance of waste management systems in Adelaide is higher compared to Stockholm. This difference is due to the virgin material recovery and energy substitution by the waste management systems. Adelaide substitutes more virgin materials than Stockholm. The overall performance of the three cities was analyzed and San Francisco was found to be top amongst the three cities.

The study aimed to develop a holistic tool for measuring the waste management performance of a city. From the study results, it is evident that San Francisco has a higher zero waste index than Adelaide and Stockholm. Virgin materials substitution, energy savings, emissions saving and water savings were also higher than the other two cities. This study was limited to the municipal waste management systems in 6 broad waste categories: paper, plastic, metal, glass, organic, and mixed municipal solid waste. Further research is required to develop a zero-waste index system for other types of waste such as commercial and institutional waste, industrial waste, and construction and demolition waste.

References

Abarca Guerrero, L., Maas, G., Hogland, W., 2012. Solid waste management challenges for cities in developing countries. Waste Manag. 33, 220–232.

Ackerman, K., Conard, M., Culligan, P., Plunz, R., Sutto, M.-P., Whittinghill, L., 2014. Sustainable food systems for future cities: the potential of urban agriculture. Econ. Soc. Rev. 45, 189–206.

Agudelo-Vera, C.M., Leduc, W.R.W., Mels, A.R., Rijnaarts, H.H.M., 2012. Harvesting urban resources towards more resilient cities. Resour. Conserv. Recycl. 64, 3–12.

Agudelo-Vera, C.M., Mels, A., Keesman, K., Rijnaarts, H., 2012. Urban harvest approach as an aid for sustainable urban resource planning. J. Ind. Ecol. 16, 839–850.

Allen, C., Gokaldas, V., Larracas, A., Ann Minot, L.A., Morin, M., Tangri, N., Tyler, B., Walker, B., 2012. On the Road to Zero Waste: Successes and Lessons from around the World. Global Anti-Incinerator Alliance, Quezon City, Philippines, pp. 1–88.

Amengual, L., 2014. 25 years of waste management in Mallorca, Balearic Islands, Spain: time to brush up on history. In: Proceedings of the Alternatives to Incineration and Landfills: Zero Waste International Alliance Conference, Nanaimo, BC, Canada, 30 July.

Anderson, C., 2011. Dealing with Zero Waste. MSW Management, Santa Barbara, CA, pp. 1–5.

Anthony, R.; Liss, G. Zero waste hierarchy: NEWS. Available online: http://zwia.org/standards/zerowaste-hierarchy/ (accessed on 8 December 2014).

Asian Development Bank, 2017. Integrated Solid Waste Management for Local Government: A Practical Guide. Asian Development Bank, Manila, Philippines.

Ayres, R.U., Ayres, L.W., 2002. A Handbook of Industrial Ecology. Edward Elgar Publishing Company, Cheltenham, UK.

Bach, B., Wilhelmer, D., Palensky, P., 2010. Smart buildings, smart cities and governing innovation in the new millennium. In: Proceedings of the 8th IEEE International Conference on Industrial Informatics, Osaka, Japan, 13–16 July, 2010, pp. 8–14.

Bartl, A., 2011. Barriers towards achieving a zero waste society. Waste Manag. 31, 2369–2370.

Bartl, A., 2014. Editorial: ways and entanglement of the waste hierarchy. Waste Manag. 34, 1–2.

Bartl, A., 2014. Moving from recycling to waste prevention: a review of barriers and enablers. Waste Manag. Res. 32, 3–18.

Bartl, A., 2014. Evolution not Revolution. Waste Management World. Available online: http://www.waste-management-world.com/articles/print/volume-14/issue-5/wmw-special-recyclingfocue/trash-talking-recycling-2035.html. (accessed on 8 December 2014).

Batty, M., Axhausen, K.W., Giannotti, F., Pozdnoukhov, A., Bazzani, A., Wachowicz, M., Ouzounis, G., Portugali, Y., 2012. Smart cities of the future. Eur. Phys. J. Spec. Top. 214, 481–518.

Benyus, J., 1997. Biomimicry: Innovation Inspired by Nature. Harper Collins Publishers Inc., New York, NY.

Bourg, D., Erkman, S., 2003. Perspectives on Industrial Ecology. Greenleaf Publishing Ltd., Sheffield, UK.

Campanelli, J., 2011. Swell idea turns sour. In: Waste & Recycling News. Vol. 17. KC Crain, Detroit, MI, p. 1.

Campbell, T., Williams, C., Ivanova, O., Garrett, B., 2011. Could 3D Printing Change the World? Technologies, Potential, and Implications of Additive Manufacturing. Retrieved from Washington DC, USA. Available online http://www.cbpp.uaa.alaska.edu/afef/Additive%20MFG%20.pdf.

Campbell-White, T., 2016. Why Zero Waste Needs To Be So Much More Than Just Zero Landfill. Available online: https://www.dssmith.com/recycling/insights/blogs/2015/12/why-zero-waste-needs-to-be-somuch-more-than-just-zero-landfill. (accessed on 14 September 2018).

Chalmin, P., Gaillochet, C., 2009. From Waste to Resources. Veolia & Cyclope, Paris, France.

Chandavarkar, N., UNDESA/UNEP, 2010. Introduction of the Report of the Secretary-General on Waste Management. UNDESA, New York, NY.

Chaudhary, H., 2013. A road map for zero waste Ahmedabad: A Visionary Document to Guide Ahmedabad towards becoming a "resource efficient and zero wastecity" by 2031. Retrieved from UNCRD/AMC, Ahmedabad, India. Available online: http://www.egovamc.com/Downloads/SWM/Ahmedabad_Road_Map_for_Zero_Waste.pdf. (accessed on 8 December 2014).

Chonga, S., Chiub, H.-L., Liaob, Y.-C., Hungc, S.-T., Pand, G.-T., 2015. Cradle to cradle®design for 3D printing. Chem. Eng. Trans. 45, 1669–1674.

City of Austin, 2008. Zero Waste Strategic Plan. City of Austin, Austin, TX.

Clay, S., Gibson, D., Ward, J., 2006. Sustainability Victoria: influencing resource use, towards zero waste and sustainable production and consumption. J. Clean. Prod. 15, 782–786.

Clift, R., 2004. Waste a waste! Mater. Today 7, 64. Available online https://www.materialstoday.com/energy/articles/s1369702104001038/. (accessed on 14 September 2018).

Cobo, S., Dominguez-Ramos, A., Irabien, A., 2018. From linear to circular integrated waste management systems: a review of methodological approaches. Resour. Conserv. Recycl. 135, 279–295.

Cohen, B., 2006. Urbanization in developing countries: current trends, future projections, and key challenges for sustainability. Technol. Soc. 28, 63–80.

Colon, M., Fawcett, B., 2006. Community based household waste management: lessons learnt from EXNORA's waste management scheme in tow south Indian cities. Habitat Int. 30, 916–931.

Connett, P., 2013. The Zero Waste Solution: Untrashing the Planet One Community at a Time. Chelsea Green Publishing, White River Junction, VT.

Cook, I.R., Swyngedouw, E., 2012. Cities, social cohesion and the environment: towards a future research agenda. Urban Stud. 49, 1959–1979.

Crocker, R., Lehmann, S., 2012. Conclusion: The culture and politics of zero waste–looking ahead. In: Crocker, R., Lehmann, S. (Eds.), Designing for Zero Waste–Consumption, Technologies and the Built Environment. Earthscan, London, UK, pp. 385–393.

Dimino, R., Warren, B., 2004. Reaching for Zero: The Citizen Plan for Zero Waste in New York City; Zero Waste Campaign. Consumer Policy Institute & Consumers Union, New York, NY, pp. 1–195.

Dobbs, R., Oppenheim, J., Thompson, F., Mareels, S., Nyquist, S., Sanghvi, S., 2013. Resource Revolution: Tracking Global Commodity Markets: Trends Survey 2013. McKinsey Global Institute, New York, NY.

D-Waste, 2013. Waste Management for Everyone. Available online http://www.d-waste.com/. (accessed on 14 September 2018).

Earth Science for Society, 2005. Megacities—Our Global Urban Future. Earth Science for Society Foundation, Leiden, The Netherlands.

Edenhofer, O., Pichs-Madruga, R., Sokona, Y., Farahani, E., Kadner, S., Seyboth, K., Adler, A., Baum, I., Brunner, S., Eickemeier, P., et al., 2014. Climate Change 2014: Mitigation of Climate Change. Contribution of Working Group III to the Fifth Assessment Report of the Intergovernmental Panel on Climate Change. Cambridge University Press, Cambridge, UK; New York, NY.

El-Chichakli, B., 2016. Five cornerstones of a global bioeconomy. Nature 535, 203–221.

Electrolux, 2014. Zero Waste: Always Trendy: Body Scan Creates Your New Wardrobe. Electrolux Design Lab, Stockholm, Sweden.

Elkington, J., 2012. The Zeronauts: Breaking the Sustainability Barrier. Routledge, New York, NY.

Ellen MacArthur Foundation, 2013. Towards the Circular Economy 1: Economic and Business Rationale for an Accelerated Transition. Ellen MacArthur Foundation, Cowes, UK.
Enkvist, P.-A., Klevnas, P., 2018. The Circular Economy–A Powerful Force for Climate Mitigation: Transformative Innovation for Prosperous and Low-Carbon Industry. Material Economics Sverige AB, Stockholm, Sweden.
ESA, 2013. 3D Printing for Space: The Additive Revolution. European Space Agency, Paris, France.
Fairburn, S., Imhof, B., Mohanty, S., 2014. Rethinking water: a CAAS (City as a spaceship) design approach. Des. J. 20, 1904–1915.
Frame, B., Brown, J., 2008. Developing post-normal technologies for sustainability. Ecol. Econ. 65, 225–241.
Funtowicz, S., Ravetz, J., 2003. Post-normal science. In: Internet Encyclopaedia of Ecological Economics. International Society for Ecological Economics, London, UK, pp. 1–10.
Garcia-Serna, J., Perez-Barrigon, L., Cocero, M.J., 2007. New trends for design towards sustainability in chemical engineering: green engineering. Chem. Eng. J. 133, 7–30.
Gharfalkar, M., Richard Court, R., Campbell, C., Ali, Z., Hillier, G., 2015. Analysis of waste hierarchy in the European waste directive 2008/98/EC. Waste Manag. 39, 305–313.
Ghisellini, P., Cialani, C., Ulgiati, S., 2016. A review on circular economy: the expected transition to a balanced interplay of environmental and economic systems. J. Clean. Prod. 114, 11–32.
Gibbs, B., 2011. Green nanotechnology. In: Mulvaney, D., Robbins, P. (Eds.), Green Technology: An A-to-Z Guide. SAGE Publications, London, UK, p. 226.
Girardet, H., 2000. Cities, people, planet. In: Urban Sustainability Schumacher Lectures. Wiley, Liverpool, UK.
Glavic, P., Lukman, R., 2007. Review of sustainability terms and their definitions. J. Clean. Prod., 1875–1885.
Graedel, T.E., 2010. Metal Stocks in Society: Scientific Synthesis. UNEP, New York, NY.
Graedel, T.E., Allenby, B.R., 2010. Industrial Ecology and Sustainable Engineering. Prentice Hall, Upper Saddle River, NJ.
Greenfield, A., Graedel, T.E., 2013. The omnivorous diet of modern technology. Resour. Recover. Conserv. 74, 1–7.
Grimm, N.B., Faeth, S.H., Golubiewski, N.E., Redman, C.L., Wu, J., Bai, X., Briggs, J.M., 2008. Global change and the ecology of cities. Science 319, 756–760.
Gutherlet, J., 2008. Recovering Resources–Recycing Citizenship: Urban Poverty Redcution in Latin Amercia. Ashgate Publishing Ltd., Aldershot, UK.
Gutherlet, J., 2010. Waste, poverty and recycling. Waste Manag. 30, 171–173.
Haas, W., Krausmann, W., Wiedenhofer, D., Markus Heinz, M., 2015. How circular is the global economy? An assessment of material flows, waste production, and recycling in the European Union and the world in 2005. J. Ind. Ecol. 19, 765–777.
Hannon, J., 2015. Waste vs. Zero Waste: The Contest for Engaging and Shaping Our Ambient "Waste-Making" Culture in Unmaking Waste 2015: Transforming Production and Consumption in Time and Place. Zero Waste SA Research Centre for Sustainable Design and Behaviour and the University of South Australia, Adelaide, Australia.
Hannon, J., Zaman, A.U., Rittl, G., Meireles, S., Demore-Palandi, F.E., 2018. Moving toward zero waste cities: a nexus for international zero waste academic collaboration (NIZAC). In: Proceedings of the 2nd Symposium on Sustainability in University Campuses (SSUC-2018), Florence, Italy, 10–12 December.
Hansen, J., 2009. Storms of my Grandchildren: The Truth about the Coming Climate Catastrophe and our Last Chanch to Save Humanity. Bloomsbury, London, UK.
Hawken, P., Lovins, A.B., Lovins, H., 1999. Natural Capitalism: Creating the Next Industrial Revolution. Little Brown and Co, Boston, MA, p. 396.
Herbert, L., 1998. The History of the Institute of Wastes Management 1898–1998: Celebrating 100 Years of Progress. Warwick Publishing Co. Ltd., England, UK.

Hestin, M., Mudgal, S., Portugal, V., Obersteiner, G., Scherhaufer, S., Beigl, P., Pertl, A., Barruetabeña, L., Salas, O., Delgado, C., et al., 2010. Zerowin: Literature Review "Approaches to Zero Waste". University of Southampton Southampton, Southampton, UK.
Hill, J., Hislop, H., Steel, C., Shaw, B., 2006. An International Survey of Zero Waste Initiatives. Green Alliance/DEFRA, London, UK.
Hodzic, A., Wiedinmyer, C., Salcedo, D., Jimenez, J.L., 2012. Impact of trash burning on air quality in Mexico City. Environ. Sci. Technol. 46, 4950–4957.
Hood, I., 2013. Ministry of Environment British Columbia. Zero Waste Business Case: Draft for Expert Review. Ministry of Environment, Vancouver, BC, Canada.
Hoornweg, D., Thomas, L., 1999. What a Waste: Solid Waste Management in Asia. East Asia and Pacific Region. World Bank, Washington, DC.
Hoornweg, D., Bhada-Tate, P., Anderson, C., 2012. What a Waste: A Global Review of Solid Waste Management. World Bank, Washington, DC.
Hoornweg, D., Bhada-Tata, P., Kennedy, C., 2014. Peak waste: when is it likely to occur? J. Ind. Ecol. 19, 117–128.
Humes, E., 2012. Garbology: Our Dirty Love Affair with Trash. Avery, New York, NY.
Hussain, C.M., 2019. Handbook of Environmental Materials Management, first ed. Springer International Publishing.
Hussain, C.M., 2020. The Handbook of Environmental Remediation: Classic and Modern Techniques. Royal Society of Chemistry.
Hussain, C.M., Keçili, R., 2019. Modern Environmental Analysis Techniques for Pollutants, first ed. Elsevier.
Hussain, C.M., Kharisov, B., 2017. Advanced Environmental Analysis-Applications of Nanomaterials. The Royal Society of Chemistry.
Hussain, C.M., Mishra, A.K., 2019. Nanotechnology in Environmental Science, 2 Volumes. John Wiley & Sons.
Inc, S.R.M.G., 2009. Environmental Life Cycle Assessment of Waste Management Strategies with a Zero Waste Objective: Study of the Solid Waste Management System in Metro Vancouver, British Columbia. Sound Resource Management Group Inc., Belkorp Environmental Services Inc & Metro Vancouver, Olympia, WA.
Isenmann, R., 2008. Further efforts to clarrify industrial ecology's hidden philosophy of nature. J. Ind. Ecol. 6, 27–48.
Iskandar, L., Tjell, J.C., 2009. Editorial–Cairo: a colossal case of waste mismanagement to learn from. Waste Manag. Res. 27, 939–940.
ISWA, 2009. Waste and Climate Change: ISWA White Paper Summary. International Solid Waste Association, Vienna, Austria.
ISWA, 2015. ISWA Annual Report. International Solid Waste Association, Vienna, Austria.
Jessen, M., 2003. Discarding the Idea of Waste: The Need for a Zero Waste Policy Now. Zero Waste Services, Nelson, BC, Canada.
Klein, J.T., 2014. Discourses of transdisciplinarity: looking back to the future. Futures 63, 68–74. Urban Sci. 2018, 2, 90 26 of 26.
Korge, K. The King's New Clothes. Available online; Yanko Design: 2014. http://www.yankodesign.com/2014/07/25/the-king%E2%80%99s-new-clothes/.
Korhonen, J., 2004. Industrial ecology in the strategic sustainable development model: strategic applications of industrial ecology. J. Clean. Prod. 12, 809–823.
Krausz, R., 2012. All for Naught? A Critical Study of Zero Waste to Landfill Initiatives. Lincoln University, Christchurch, New Zealand.
Krausz, R., Hughey, K.F.D., Montgomery, R., 2013. Zero waste to landfill: an unacknowledged supermegaproject. Linc. Plan. Rev. 5, 10–26.
Krzeminska, A.E., Zareba, A.D., Dzikowska, A., Jarosz, K.R., 2017. Cities of the future-bionic systems of newurban environment. Environ. Sci. Pollut. Res., 1–9.

Lehmann, S., 2010. The Principles of Green Urbanism: Transforming the City for Sustainability, first ed. Earthscan, Oxford, UK.
Lehmann, S., 2010. Resource recovery and materials flow in the City: zero Waste and sustainable consumption as paradigms in urban development. Sustain. Dev. Law Policy Sustain. Dev. Urban Environ. 11, 28–68.
Lehmann, S., 2010. Green urbanism: formulating a series of holistic principles. Sapiens 3, 2.
Lehmann, S., 2011. Optimizing urban material flows and waste streams in urban development through principles of zero waste and sustainable consumption. Sustainability 3, 155–183.
Lehmann, S., 2011. What is green urbanism? Holistic principles to transform cities for sustainability. In: Blanco, J. (Ed.), Climate Change–Research and Technology for Adaptation and Mitigation. InTech, Rijeka, Croatia.
Lehmann, S., Crocker, R., 2012. Designing for Zero Waste: Consumption, Technologies and the Built Environment; Earthscan Series on Sustainable Design. Earthscan, London, UK.
Lenton, T.M., Ciscar, J., 2013. Integrating tipping points into climate impact assessments. Clim. Chang. 117, 585.
Leroux, K., 2001. Is zero waste possible? Waste Age 32, 102.
Levin, K., Cashmore, B., Bernstien, S., 2009. Playing it forward: path dependency, progressive incrementalism, and the "super wicked" problem of global climate change. In: Climate Change: Global Risks, Challenges and Decisions Congress. International Nuclear Information System, Copenhagen, Denmark.
Levin, K., Cashore, B., Bernstein, S., Auld, G., 2012. Overcoming the tragedy of super wicked problems: constraining our future selves to ameliorate global climate change. Policy Sci 45, 123–152.
Levitzke, V., 2012. Foreword: designing for zero waste. In: Crocker, R., Lehmann, S. (Eds.), Designing for Zero Waste: Consumption, Technologies and the Built Environment. Earthscan, London, UK.
Loiseau, E., Saikku, L., Antikainen, R., Droste, N., Hansjürgens, B., Pitkeanen, K., Leskinen, P., Kuikman, P., Thomsen, M., 2016. Green economy and related concepts: An overview. J. Clean. Prod. 139, 361–371. Urban Sci. 2018, 2, 90 22 of 26.
Lombardi, E., 2001. Beyond recycling! Zero waster darn near. Biocycle 42, 75.
Lombardi, E., Bailey, K., 2015. The Community Zero Waste Roadmap. Eco-Cycle Soutions, Bolder, CO.
Mace, M., Szaky, T., 2016. TerraCycle: Recycling Won't Solve the Waste Dilemma, but Consumers Can. Edie.Net. Available online: https://www.edie.net/news/5/We-can-t-recycle-our-way-out-of-the-waste-crisis-says-TerraCycle-founder/.
Maciejczak, M., 2015. How to analyse the bioeconomy. Ann. Pol. Assoc. Agric. Econ. Agribus XVI, 165–171.
Manahan, S.E., 1999. Industrial Ecology: Environmental Chemistry and Hazardous Waste. Lewis Publishers, Boca Raton, FL.
Marshall, R.E., Farahbakhsh, K., 2013. Systems approaches to integrated solid waste management in developing countries. Waste Manag. 33, 998–1003.
Mavropoulos, A., 2010. Globalization, Megacities and Waste Management. ISWA–White Paper, Athens, Greece.
Mavropoulos, A., 2010. Waste Management 2030+. Waste Management World. Available online: http://www.waste-management-world.com/articles/print/volume-11/issue-2/features/waste-management-2030.html.
Mavropoulos, A., 2014. Megacities Sustainable Development and Waste Management in the 21st Century. ISWA, Vienna, Austria.
Mavropoulos, A., Newman, D., ISWA, 2015. Wasted Health: The Tragic Case of Dumpsites. International Solid Waste Association, Vienna, Austria.

Mavropoulos, A., Chohen, P., Greedy, P., Plimakis, S., Marinheiro, L., Law, J., Loureiro, A., 2017. A Roadmap for Closing Waste Dumpsites: The World's Most Polluted Places. International Solid Waste Association (ISWA), Vienna, Austria.
Mezher, T., 2011. Building future sustainable cities: the need for a new mindset. Constr. Innov. 11, 136–141.
Morgan, J., 2013. Amaze project aims to take 3D printing "into metal age". In: BBC News: Science and Environment. BBC News, Science and Environment, London, UK.
Murphy, S., Princetl, S., 2013. Zero waste in Los Angeles: is the emperor wearing any clothes? Resour. Conserv. Recycl. 81, 40–51.
Murray, R., 1999. Creating Wealth from Waste. Demos, London, UK.
Murray, R., 2002. Zero Waste. Greenpeace Environmental Trust, London, UK.
Nader, S., 2009. Paths to a low carbon economy–the Masdar example. Energy Procedia 1, 3951–3958.
Nanotechnology Development Blog, 2013. Green Refineries Will Offer Zero Waste Products. Available online: http://www.nanotechnologydevelopment.com/news/green-refineries-will-off-zero-wasteproducts.html.
Naphade, M., Banavar, G., Harrison, C., Paraszczak, J., 2011. Smarter cities and their innovation challenges. Computer 44, 32–39.
Oyoo, R., Leemans, R., Mol, A.P.J., 2011. Future projections of urban waste flows aand their impacts in African metropolises cities. Int. J. Environ. Res. 5, 705–724. Urban Sci. 2018, *2*, 90 19–26.
Palmer, P., 2013. The Faux Zero Movement Is Spreading. Zero Waste Institute. Available online: http://zerowasteinstitute.org/wp-content/uploads/2013/03/Faux-Zero-Waste.pdf. (accessed on 8 December 2014).
Pfau, S.F., Hagens, J.E., Dankbaar, B., Smits, A.J.M., 2014. Visions of sustainability in bioeconomy research. Sustainability 6, 1222–1249.
Pietzsch, N., Ribeiro, J.L.D., De Medeiros, J.F., 2017. Benefits, challenges and critical factors of success for zero Waste: a systematic literature review. Waste Manag. 67, 324–353.
Pollans, L.B., 2017. Trapped in trash: "modes of governing" and barriers to transitioning to sustainable waste management. Environ. Plan. 49, 2300–2323.
Pongrácz, E., Pohjola, V.J., 2004. Re-defining waste, the concept of ownership and the role of waste management. Resour. Conserv. Recycl. 20, 141–153.
Porritt, J., 2007. Capitalism as if the World Matters, Revised ed. Earthscan, London, UK.
Porse, E.C., 2013. Stormwater governance and future cities. Water 5, 29–52.
Premalatha, M., Tauseef, S.M., Abbasi, T., Abbasi, S.A., 2013. The promise and the performance of the world *is* first two zero carbon eco-cities. Renew. Sust. Energ. Rev. 25, 660–669.
Ragossnig, A., 2006. Zero waste—a desirable goal in residue utilization? (editorial). Waste Manag. Res. 24, 403.
Riffat, S., Powell, R., Aydin, D., 2016. Future cities and environmental sustainability. Future Cities Environ. 2, 1–2.
Ryan, P.G., Moore, C.J., van Franeker, J.A., Moloney, C.L., 2009. Monitoring the abundance of plastic debris in the marine environment. Philos. Trans. R. Soc. B, 1999–2012.
Schneider, S., 2009. The worst-case scenario. Nature 458, 1104–1105.
Seadon, J.K., 2010. Sustainable waste management systems. J. Clean. Prod., 1639–1651.
Seife, C., 2000. Zero: The Biography of a Dangerous Idea. Penguin Group, New York, NY.
Silva, A., Rosano, M., Stocker, L., Gorissen, L., 2017. From waste to sustainable materials management: three case studies of the transition journey. Waste Manag. 61, 547–557.
Snow, W., Dickinson, J., 2001. The End of Waste: Zero Waste by 2020. Zero Waste New Zealand Trust, Auckland, New Zealand.
Snyman, J., Vorster, K., 2010. Towards zero waste: a case study in the City of Tshwane. Waste Manag. Res. 29, 512–520.

Stern, N., 2009. A Blueprint for a Safer Planet: How to Manage Climate Change and Create a New Era of Progress and Prosperity. The Bodley Head, London, UK.

Swapan, M.S.H., Zaman, A.U., Ahsan, T., Ahmed, F., 2017. Transforming urban dichotomies and challenges of south Asian megacities: rethinking sustainable growth of Dhaka, Bangladesh. Urban Sci. 1, 31.

Taylor, S., 2014. Waste Management Implications of 3D Printing. EcoMENA, Doha, Qatar.

Themelis, N.J., 2009. Don't trash waste-to-energy plants. In: Cape Cod Times. Cape Cod Media Group, Hyannis, MA, p. 149. ISWA. ISWA's Mission. Available online http://www.iswa.org/iswa/organisation/about-iswa/. (accessed on 14 September 2018).

Themelis, N.J., n.d. Search Result: Comments by Prof. Themelis on "Zero Waste" Discussion in D-Waste Blog. Available online: http://www.seas.columbia.edu/earth/wtert/ (Accessed 8 December 2014).

Thompson, A., 2014. Burning trash bad for humans and global warming. Sci. Am. Available online: http://www.scientificamerican.com/article/burning-trash-bad-for-humans-and-globalwarming/.

Troschinetz, A.M., Mihelcic, J.R., 2009. Sustainable recycling of municipal solid waste in developing countries. Waste Manag., 915–923.

Trotti, J., 2012. MSW Management: Swana Golden Anniversary Issue. Vol. 22 Forrester Media Inc, Sanata Barbara, CA, pp. 1–104.

Tsuzuki, T., 2010. Nanotechnology Commercialization. CRC Press, Boca Raton, FL.

UNEP, UNITAR, Hyman, M., Turner, B., Carpintero, A., 2013. Guidelines for National Waste Management Strategies: Moving from Challenges to Opportunities. United Nations Environment Programme (UNEP) & United Nations Institute for Training and Research (UNITAR) & Inter-Organisation for the Sound Management of Chemicals (IOMC), Osaka, Japan.

UNFPA, 2007. State of World Population 2007: Unleashing the Potential of Urban Growth. UNFPA, New York, NY.

UN-HABITAT, 2008. State of the World's Cities 2010/2011: Bridging the Urban Divide, first ed. Earthscan, Oxford, UK.

UN-Habitat, 2010. Solid Waste Management in the Worlds 0s Cities. Earthscan, London, UK.

United Nations Economic Commission for Europe (UNECE), 2011. Climate Neutral Cities: How to Make Cities Less Energy and Carbon Intensive and More Resilient to Climatic Challenges [Online]. Available. http://www.unece.org/fileadmin/DAM/hlm/documents/Publications/climate.neutral.cities_e.pdf. (accessed 18.07.12.).

USEPA. Solid Waste Management Hierarchy. Available online. 2013; http://www.epa.gov/wastes/nonhaz/municipal/hierarchy.htm.

Van Ewijk, S., Stegemann, J.A., 2016. Limitations of the waste hierarchy for achieving absolute reductions in material throughput. J. Clean. Prod. 132, 122–128.

Van Vliet, A., 2014. Zero Waste Europe Case Study 1: The Story of Capannori. Zero Waste Europe, Leiden, The Netherlands. Urban Sci. 2018, 2, 90 24 of 26.

Van Vliet, A., 2014. Zero Waste Europe Case Study 2: The Story of Argentona. Zero Waste Europe, Leiden, The Netherlands.

Van Vliet, A., 2014. Zero Waste Europe Case Study 3: The Story of Vrhinika "Slovenian Traiblazers". Zero Waste Europe, Leiden, The Netherlands.

Veleva, V., Bodkin, G., Todorova, S., 2016. The need for better measurement and employee engagement to advance a circular economy: Lessons from Biogen's "zero waste" journey. J. Clean. Prod. 154, 517–529.

Velis, C.A., Wilson, D.C., Rocca, O., Smith, S.R., Mavropoulos, A., Cheeseman, C.R., 2012. An analytical framework and tool ("InteRa") for integrating the informal recycling sector in waste and resource management systems in developing countries. Waste Manag. Res. 30, 43–66.

Verstraete, W., Vlaeminck, S.E., 2011. ZeroWasteWater: short-cycling of wastewater resources for sustainable cities of the future. Int. J. Sustain. Dev. World Ecol. 18, 253–264.
Vezzoli, C., Manzini, E., 2008. Design for Environmental Sustainability. Springer-Verlag London Limited, London, UK.
Wainwright, O., 2014. Work begins on the world's first 3D-printed house. In: The Guardian: Architecture and Design Blog. The Guardian, Architecture and Design Blog, London, UK.
Waste Watch, UK, 2004. History of Waste and Recycling, in Wasteonline Information Sheet. Waste Watch UK, London, UK.
WasteMINZ, Eunomia Research & Consulting, 2018. Rebooting Recycling—What can Aotearoa do? A Discussion Paper Presented by the Waste Management Institute of New Zealand (WasteMINZ). Eunomia Research & Consulting, Auckland, New Zealand.
Wiedinmyer, C., Yokelson, R.J., Gullett, B.K., 2014. Global emissions of trace gases, particulate matter, and hazardous air pollutants from open burning of domestic waste. Environ. Sci. Technol. 48, 9223–9530.
Williams, I., 2014. ZEROWIN. Southampton University, Southampton, UK. Available online: http://www.southampton.ac.uk/mediacentre/news/2013/feb/13_30.shtml. (accessed on 8 December 2014).
Wilson, D.C., 2007. Development drivers for waste management. Waste Manag. Res. 25, 198–207.
Wilson, D.C., Velis, C., Cheeseman, C., 2006. Role of informal sector recycling in waste management in developing countries. Habitat. Int. 30, 797–808.
Wilson, D.C., Araba, A.O., Chinwah, K., Cheeseman, C.R., 2009. Building recycling rates through the informal sector. Waste Manag. 29, 629–635.
Wilson, D.C., Rodic, L., Modak, P., Soos, R., Carpintero, A., Velis, C., Iyer, M., Simonett, O., 2015. Global Waste Management Outlook. ISWA & UNEP, Vienna, Austria.
Wilson, D.C., Rodic, L., Modak, P., Soos, R., Carpintero, A., Velis, C., Iyer, M., Simonett, O., 2015. Global Waste Management Outlook: Summary for Decision-Makers. ISWA & UNEP, Vienna, Austria.
WMAA, AIEN, 2013. Resource Recovery–It's just good business. In: Proceedings of the 4th Australasian Industrial Ecology Conference, Ulsan, Korea, 25–28 June.
Young, C.-Y., Ni, S.-P., Fan, K.-S., 2010. Working towards a zero waste environment in Taiwan. Waste Manag. Res. 28, 236–244.
Zaman, A.U., Lehmann, S., 2011. Challenges and opportunities in transforming a City into a "zero Waste City". Challenges 2, 73–93.
Zaman, A.U., Lehmann, S., 2011. Urban growth and waste management optimization towards "zero waste city". City Cult. Soc. 2, 177–187.
Zero Waste Canada. Programme for the ZWIA14 dialogue. Available online: http://www.zerowastecanada.ca/ (accessed on 8 December 2014).
Zero Waste Europe, 2017. The Zero Waste Masterplan: Start-Up Toolkit for European City Planner Policy-Makers and Community Leaders. Zero Waste Europe, Ixelles, Belgium.
ZeroWIN. Zero Waste Industrial Networks Project. Available online: http://www.zerowin.eu/ (accessed on 8 December 2014).
Zotos, G., Karagiannidis, A., Zampetoglou, S., Malamakis, A., Antonopoulos, I.S., Kontogianni, S., Tchobanoglous, G., 2008. Developing a holistic strategy for integrated waste management within municipal planning: challenges, policies, solutions and perspectives for Hellenic municipalities in the zero-waste, low-cost direction. Waste Manag. 29, 1686–1692.

ZWIA, 2004. Zero Waste Definition Adopted by Zero Waste Planning Group [Online]. Available: http://www.zwia.org/main/index.php?option¼com_content&view¼article&id¼49&Itemid¼37 (accessed 16.07.2010).
ZWIA, 2009. Global Principles for Zero Waste Communities. Zero Waste International Alliance (ZWIA), Brighton, UK.

Further reading

Auckland Council, 2012. Auckland Waste Management and Minimisation Plan: Getting Auckland Is Waste Sorted. Auckland Council, Auckland, New Zealand.
Australian Living Lab Innovation Network, Expression of Interest. Available online. https://openlivinglabs.net.au/. (Accessed 8 December 2014).
Carrinho, M. The eco Industrial Park (EIP) of Chamusca. Available online: http://ecoteste.blogspot.co.nz/2007/10/eco-industrial-park-eip-of-chamusca.html (accessed 8 December 2014).
Chesbrough, H., 2003. Open Innovation. The New Imperative for Creating and Profiting from Technology. Harvard Business School Press, Boston, MA.
Curran, T., Williams, I.D., 2012. A zerowaste vison for industrial networks in Europe. J. Hazard. Mater. 207, 3–7.
Hannon, J., 2018. Changing Behaviour: (New Zealand's Delay & Dysfunction in Utilising) Economic Instruments in the Management of Waste? New Zealand Product Stewardship Council, Palmerston North, New Zealand.
Hogg, D., Ballinger, A., 2015. The Potential Contribution of Waste Management to a Low Carbon Economy; Zero Waste Europe, Zero Waste France. ACR+, Eunomia Research & Consulting, Brussels, Belgium.
Homer-Dixon, T., 2006. The Upside of Down: Catastrophe, Creativity and the Renewable of Civilistion. Text Publishing, Melborne, Australia.
ISWA, ABRELPE, 2003. Solid Waste: Guidelines for Successful Planning. ISWA & ABRELPE, Sao Paulo, Brazil.
McDonough, W., Braungart, M., 2002. Cradle to Cradle: Remaking the Way we Make Things. North Point Press, New York, NY.
Moore, C.J., 2008. Synthetic polymers in the marine environment: a rapidly increasing, long-term threat. Environ. Res., 131–139.
Niitamo, V.-P.; Kulkki, S.; Eriksson, M.; Hribernik, K.A. State-of-the-Art and good practice in the field of living labs. In Proceedings of the 12th International Conference on Concurrent Enterprising: Innovative Products and Services through Collaborative Networks, Milan, Italy, 20 November 2015; pp. 349–357.
Phillips, P.S., Tudor, T., Bird, H., Bates, M., 2011. A critical review of a key waste strategy initiative in England: zero waste places projects 2008–2009. Resour. Conserv. Theatr. Rec. 55, 335–343.
Rodrigues, J. ECO LivingLab@Chamusca. Available online: http://www.openlivinglabs.eu/livinglab/ecolivinglabchamusca (accessed 14 September 2018).
Schumacher, J., 2013. Alcotra Innovation Project: Living Labs Definition, Harmonization Cube Indicators & Good Practices. Alcotra, New York, NY.
Simon, J.M. The First European Zero Waste Research Center—Capannori, Italy. Available online: http://www.zerowasteeurope.eu/2011/05/the-first-european-zero-waste-research-center-capannori-italy/ (accessed 14 September 2018).
Snow, W., Dickinson, J., 2003. Getting There: The Road to Zero Waste: Strategies for Sustainable Communities. The Zero Waste New Zealand Trust and Envison New Zealand Ltd., Auckland, New Zealand.

UCLG Commeetee. Capannori, Italy: The First Case of the Application of the:"Zero Waste strategy" in Italy. Available online: https://www.uclg-cisdp.org/sites/default/files/Capannori_2010_en_FINAL.pdf (accessed on 14 September 2018).
UNEP, 2009. Developing Integrated Solid Waste Plan: ISWM Plan–Training Manuals Volume 1–4. United Nations Environmental Programme Division of Technology, Industry and Economics, International Environmental Technology Centre, Osaka/Shiga, Japan.
University of South Australia, China-Australia Centre for Sustainable Urban Development. Available online. http://www.unisa.edu.au/research/zerowastecentre/. (Accessed 8 December 2014).
Venn, W., Living Labs to Stimulate Low Carbon Living. Available online. http://w3.unisa.edu.au/unisanews/2015/April/story8.asp. (Accessed 8 December 2014).
Vliet, A.V. n.d. The Story of Capannori—A Zero Waste Champion. Available online: https://zerowasteeurope.eu/2013/09/the-story-of-capannori-a-zero-waste-champion/ (accessed on 14 September 2018). Urban Sci. 2018, *2*, 90 21 of 26.
Wilson, R., Ryan, A., Living Lab a Fresh Approach to Sustainable Research. Available online. http://www.massey.ac.nz/news/?id=5161/. (Accessed 8 December 2014).
Zero Waste Italy, The open case study of the Zero Waste Research Center of Capannori (Lucca) on the Lavazza coffee pods starts delivering results. Available online. http://www.zerowasteitaly.org/the-opencase-study-of-the-zero-waste-research-center-of-capannori-lucca-on-the-lavazza-coffee-pods-startsdelivering-results/. (Accessed 14 September 2018).

CHAPTER NINE

Toward "Zero Liquid Discharge" industrial facilities: Reducing the impact on freshwater resources by reusing industrial and urban wastewaters

Antonio Santos Sánchez
Federal University of Ouro Preto, Ouro Preto, Brazil

Contents

1 An introduction to zero liquid discharge 215
2 Reuse of municipal wastewater 217
3 A successful case study: Aquapolo reclamation plant 220
4 Outcomes of the Aquapolo project 232
5 How the Aquapolo project helped fostering water efficiency and internal reuse projects among the participating companies 235
6 Conclusions 242
References 243

1 An introduction to zero liquid discharge

Zero liquid discharge (ZLD) consists of the conversion of wastewater to recycled water and solids, thus limiting the liquid waste at the end of an industrial process to zero.

A ZLD treatment system combines clarifiers, membrane technologies such as ultrafiltration and reverse osmosis, with final stages of brine concentrators, evaporators and crystallizers (Hussain and Kharisov, 2017; Hussain and Mishra, 2019; Hussain, 2019). While recovering the greatest part of the liquid waste for reuse, it can also retrieve valuable byproducts from the waste effluents. In addition, ZLD systems produce a dry, solid cake for disposal, which is useful to concentrate any potentially hazardous compounds that otherwise would be diluted in a liquid stream.

A ZLD process can mitigate two problems:

Concepts of Advanced Zero Waste Tools
https://doi.org/10.1016/B978-0-12-822183-9.00009-X

- Most industrial processes produce waste effluents that need to be treated to preserve the environment and complain with regulations (Hussain and Keçili, 2019; Hussain, 2020). Urban areas are subjected to the impact of both industrial and domestic wastewaters and have to deal with an increasingly polluted aquatic environment. An industrial facility may be located on a site that is subjected to severe environmental regulations or has high discharge fees.
- Water scarcity is already a reality in many urban areas and that problem will get worse: it is expected that two-thirds of the global population may live in regions with limited access to freshwater resources by 2050 (UN DESA, 2018). A ZLD treatment system typically recovers around 95% of the liquid waste for reuse (SAMCO Tech, 2019). Despite having high initial costs, implementing a ZLD system in an industrial facility can be a strategic move in the context of rising water costs and recurrent droughts.

Industrial effluents and brines are typically discharged into the environment after treatment but, instead, can undergo further tertiary treatment that polishes them until becoming suitable for uses as industrial water or for irrigation.

Produced water from the oil and gas industry is an example of a hazardous effluent that is already been treated to high-quality standards. In the San Ardo reclamation facility (California), produced water undergoes a multistep treatment process involving oil/water separation, chemical treatment, and reverse osmosis to remove oil, solids, and salts. Then a portion of the treated effluent is used as cooling tower water, while the rest undergoes further conditioning steps that include ammonia removal, pH adjustment, ion balancing, and wetlands to create water suitable to recharge a shallow aquifer that is used in the area for crop irrigation (Myers, 2014).

Another interesting example is the brine from inland desalination facilities that treat brackish waters. That polluting effluent can be used as a resource to irrigate halophytes or produce algae and fish, as confirmed by examples from arid regions around the world (Sánchez et al., 2015a; Sánchez and Matos, 2018). At the same time, the EU-funded "Arava" project researches the use of algae and bacteria combinations to treat saline wastewater from the food industry (aquaculture), which is rich in nitrogen and phosphate (Saltgae Project, 2016).

The production of recycled water from domestic and industrial effluents is one of the most interesting applications of ZLD treatment systems. However, these technologies also allow recovering other valuable resources in the form of salts and minerals. The EU-funded "Zero Brine" project goes in that direction. It aims the recovery of minerals such as magnesium from brine generated by process industries. Since 2017, this project has pilot plants operating in a port facility, a coal plant, a silica plant, and a textile factory. The system proposed by Zero Brine consists of nanofiltration followed by an evaporator and crystallizer (Zero brine Project, 2017).

This chapter focuses on the opportunities to reduce the impact that industrial facilities have on freshwater resources using two strategies. The first strategy is recycling municipal wastewaters through tertiary treatment and using it as an input for industrial processes. The second strategy is to achieve ZLD by treating the effluents generated at the facility (domestic, industrial, and also stormwater collected at the site) separately and then reuse the effluents as industrial water. By these means, the industry can significantly reduce the withdrawal of fresh water. To illustrate this, the case study of a petrochemical complex supplied by recycled municipal wastewater will be assessed, and then the case of a ZLD chemical plant within the complex will be looked at in more detail.

The chemical industry is highly water-intensive: water is not only used for cooling purposes (around 90%) but also as raw material, in cleaning, transport, as a solvent and as part of the final product. The sector also faces the risk of spillage of hazardous chemicals which can affect the quality of local water resources (CDP Global, 2019). The assessed ZLD facility managed to treat and reuse all its effluents, so all the water input is either incorporated in the product or evaporated. As a consequence of both using municipal wastewater as the main water input and reusing its own effluents, this facility achieved values of the "fresh water index" (m^3 of fresh water needed for one ton of product) that can be considered as a benchmark for the chemical industry.

2 Reuse of municipal wastewater

Water supply security greatly depends on a combination of increasing the efficiency in water usage at urban, agricultural, and industrial levels and, in parallel, considering "alternative" water sources. Those sources mainly consist of rainwater harvesting, desalination, and the reuse of urban and industrial wastewater (Sánchez et al., 2015b; Sánchez and Matos, 2018; Oliveira-Esquerre et al., 2009). They can replace "traditional" water sources (surface and ground fresh water) in some uses, even as drinking water if properly treated. Traditionally, water systems have been based on the extraction of the water available in the local watershed. Eventually, as the local water demand rises and nearby water sources become polluted or depleted, the water intake has to move farther or deeper. After the extraction or the import from other regions, that water has to be treated to drinkable standards and then distributed throughout the network. After its use, the water loses its potable condition regardless of its level of contamination and has to go through the sewer system to the wastewater treatment plant before been discharged back to the watershed. Nowadays, society is questioning that water scheme. For example, an increasing number of citizens prefer living and working in buildings that replace the use of high-quality drinkable water for toilet flushing or gardening with low treated, reused graywater, or rainwater. However, at an urban level, it is still soon to see a large-scale network that collects and distributes non-drinkable water to supply non-drinkable uses.

Megacities, such as Delhi and Mexico City, are already using reused urban wastewater for irrigation and for cooling of power plants. Those two cities bring interesting study cases: they have the potential for increasing wastewater reuse in the following years and have realized its importance as a strategic source. In the case of Delhi, it faces an urgent water crisis due to the increase in demand and the pollution of its traditional sources, such as the Yamuna and Ganga rivers. To solve Delhi's water supply challenge, "traditional" water sources such as groundwater extraction and new dams in the Himalayas are being considered. Currently, the Delhi Jal Board supplies daily more than 416 million liters of treated wastewater to irrigation in parks and for cooling in power plants, and there are several small reuse projects on its way (Government of NCT of Delhi, 2009).

In México City, the reuse of treated wastewater for non-drinkable purposes already supplies 7% of all water consumption. It can be considered as one of the world's largest wastewater irrigation systems (Siemens et al., 2008; Seguí Amórtegui et al., 2019). The metropolitan area of the Valley of México has a population of 22 million people. Its water consumption is due to 48.4% agriculture, 45.3% domestic, 4.5% industry, and 1.8% for cooling thermal power plants. Therefore, half of its water consumption refers to non-drinkable uses and could be supplied by reclaimed wastewater. As the Valley of Mexico's main water source is groundwater, the growing demand pushes the local aquifers to the limit, so the supply has to be complemented with water pumped from nearby regions.

Small-scale water reclamation plants are also an attractive possibility for rural communities and urban districts. After analyzing the challenges faced by expanding cities in Africa, some authors have concluded that decentralized and modular solutions that combine surface water, groundwater, and graywater recycling will bring more flexibility and adaptation (Jacobsen et al., 2013). The African city of Windhoek, located in an arid region of Namibia, is one of the finest examples of that 26% of the city's water supply already comes from wastewater reuse. The city has two relatively small reclaimed water plants, one of which is used to irrigate parks, golf courses, and cemeteries while the other contributes to the potable water consumption of the city being one of the few systems in the world that recycles treated wastewater for drinking water (Lahnsteiner and Lempert, 2007; van der Merwe, 2009).

At the same time, industry makes use of reclaimed wastewater more and more frequently as a basic measure to lower costs and comply with environmental requirements, while irrigated agriculture is starting to see treated wastewater as a key strategic resource (Vigneswaran, 2009). In countries with a desert climate, wastewater reuse is growing due to having better energy efficiency and thus lesser costs than desalination (Consortium CH2M Hill BV, 2016).

Nice examples of reuse of urban and industrial wastewaters can be found around the world, with large-scale examples such as Morigasaki in Tokyo (Japan), Huai Fang in Beijing (China), Changi (Singapore), Mafraq (Jordania), the Shafdan plant in the Dan region (Israel), Aquapolo in São Paulo (Brazil), Kwinana in Perth (Australia), or the Stickney (Illinois) and Orange County (California) water reclamation plants in the USA, just to name the largest ones (NRC, 2012; Lazarova et al., 2013; WEF, 2018).

To develop a successful water reclamation scheme, it is key to identify and partner with potential consumers that may use recycled water. Stable demand for recycled water will make economically viable building the infrastructure for its continuous supply. Sometimes the potential consumers may be located at significant distances, such as in the case of Shafdan. That plant treats wastewater form Tel Aviv urban region (2.5 million people) and diverts it for agricultural use in Southern Israel, over 100 km away. The cities where the effluent is extracted from succeeded in partner with farmers in the Negev Desert, which have to irrigate their crops in a very arid environment. Shafdan system consists of a preliminary treatment of the collected wastewater followed by its injection into an aquifer where the recycled water is naturally filtrated in the sands and stored with minimal evaporation losses, and finally, a pumped distribution system to transport the abstracted effluent to the consumers in the desert. Recycled water is sold at a discount rate between 30% and 50% in comparison with fresh water, thanks to a subsidy from the Ministry of Finance to cover the difference between supply costs and tariffs. In addition, Shafdan has started to produce 2000 tons of fertilizer per year, mostly from recovered phosphorous, which is marketed and sold to an established network of blenders and distributors. The success of the Shafdan facility helped pave the way for a national strategy in effluent reuse, and nowadays, 40% of Israel's agricultural needs come from recycled wastewater.

Shafdan is an example for other urban areas worldwide that could develop a similar scheme of wastewater reuse but hasn't done it so far, even when located in regions with water scarcity. Los Angeles metropolitan area, with a population of 13.1 million people, is one of them. Los Angeles has four water reclamation plants: Hyperion, Terminal Island, Donal C. Tillman, and Glendale. After treating the municipal wastewater, most of the flow is discharged into the Pacific Ocean, nearby lakes, rivers, or used for groundwater replenishment. As there isn't a network that supplies that treated wastewater to potential customers, recycling provides currently only 2% of Los Angeles' water. The city's water utility offers free disinfected recycled water for some approved purposes (industrial, landscape, and recreational purposes) so customers interested in getting free recycled water still can bring their own water containers to a recycled water fill station at Glendale's water reclamation plant. There is an ambitious goal of recycling all the treated wastewater produced by 2035: the city wants to upgrade Hyperion Water Reclamation Plant with advanced treatment technology, close the outfall that discharges wastewater into the ocean, and pipe the recycled water inland to inject it into the aquifers that underlie much of the L.A. Basin.

In the case of Shaftan as well as in Los Angeles' project, the key element is the existence of an underground layer of water-bearing permeable rock that filters the treated wastewater and stores it so groundwater can be later extracted using a water well. However, wastewater can also be treated and then supplied directly through pipes to consumers, without previous storage in an aquifer.

3 A successful case study: Aquapolo reclamation plant

Hereafter, it is presented the case of the Aquapolo wastewater reclamation plant, which treats wastewater from a heavy populated area in São Paulo (Brazil) and delivers it, directly, to large industrial consumers through a distribution network that supplies 650 L/s of recycled water for non-drinkable purposes.

Aquapolo started its operation in 2012, and it's the largest of its kind in the Southern Hemisphere. It was born in 2009 from a joint venture between the city's water utility (Sabesp, owning 49%) and a private engineering company (Foz do Brasil/Odebretch Ambiental, later bought by Brookfield and renamed BRK Ambiental, owning 51%). The city's water utility provided access to the municipal sewage and the private company invested in building the reclamation plant and the pipes to supply recycled water to a petrochemical complex located 17 km away from the plant. Chemical industries at the complex were large consumers of fresh water, which was used for non-drinkable purposes such as cooling towers and washing boilers and other equipment. Their water sources were the Tamanduateí river and other nearby sources (380 L/s) and fresh water from the city's utility (191 L/s) (Ambiental, 2013). When compared to drinkable water from the mains, the cost of supplied recycled water ranges between 3 and 10 times cheaper (Sabesp, 2019). Aquapolo created a win–win situation for all the parties: the municipal utility, the engineering company that would operate the plant and the consumers at the complex. A 41 year's contract was signed between the parties, and the investment from the private engineering company came soon after. Table 1 shows the main data of the project.

Aquapolo is located in São Paulo, a densely populated area with 21 million inhabitants. This region suffered from a severe drought between 2013 and 2015, and it was during that crisis when this reclamation plant shined as one of the city's most valuable water infrastructure assets. The 2013–2015 Drought threatened the economy of the city, which is strongly based on commerce and industry. To prevent rationing, both the population and the industry were called to make rational use of water and avoid any waste of the resource. Overall, São Paulo managed to reduce its per capita

Table 1 Aquapolo in numbers.

Extension of the main pipe from plant to customers	17 km
Extension of the distribution network	4.2 km
Built area in the reclamation plant	15,000 m^2
Storage capacity - distribution water tanks	70,000 m^3
Plant capacity	1000 L/s
Current flow	650 L/s
Volume of fresh water that is saved	2.58 million m^3 per month
Lifespan	41 years (beginning in 2012)

consumption of water by 25%. The city's water utility succeeded in reducing water losses in the city mains by 5% through pressure reduction and proactive leak detection programs (Ministry of the Cities of Brazil, 2018). As the population became aware of the problem many households, hotels, hospitals, and shopping centers invested in water-saving devices for bathrooms as well as in rainwater harvesting and reuse systems. For the large water consumers in the industrial sector, the threat of water rationing due to the shortage implies in the possibility of considerable financial losses.

That was the case of Braskem, which consumes most of the recycled water produced at Aquapolo. Thanks to the continuous supply of recycled water, that company avoided having its operations halted in 2015 which would have resulted in 50 million USD of financial losses (Braskem Corp, n.d.).

Braskem is the Americas' top thermoplastic resins producer and one of the main players behind the Aquapolo project. During São Paulo's drought, water consumption in the company was six times lesser than the average of the international chemical industry, according to data from the International Council of Chemical Associations (ICCA). Braskem's water consumption index in 2014 was 2.16 m^3 per ton of product while the average of the World's chemical industry was at 25.90 m^3 per ton of product (ICCA, 2015). The following years saw a worsening in Braskem's water consumption index which rose up to 4.28 m^3 per ton of product in 2018 (Braskem Corp, n.d.).

The company operates four large petrochemical complexes in Brazil (São Paulo, Camaçari, Triunfo and Duque de Caxias). Its basic petrochemical units produce compounds such as ethylene, propylene, benzene, butadiene, toluene, xylene, and isoprene. These chemical products are supplied to the company's nearby polymer units as well as to other chemical companies based within the same complexes. Overall, 26.3% of the consumed water in Braskem's processes came from reclaimed effluents such as recycled wastewater or process water. In São Paulo – the petrochemical complex that is supplied by Aquapolo – that index is already 97%.

In 2012, while Aquapolo was being inaugurated, Braskem started up another major project of effluent reuse. It was in the Camaçari unit where it partnered with CETREL, a private company in charge of the environmental management of the petrochemical complex. Camaçari's complex is located in a very humid region that has average precipitation values of 2000 mm evenly distributed throughout the year. Owing to that particularity, this project was designed to collect the stormwater from large areas such as containment basins around petrochemical storage tanks and then mix it with selected process effluents that have low organic content. The resulting mixed wastewater (60% stormwater and 40% industrial effluent) ends in an open reservoir with 1,600,000 m^3 of capacity. From the reservoir's surface, the wastewater is pumped to a reclamation plant where it undergoes coagulation, flocculation, and dissolved air floatation. The resulting recycled water is mostly used as cooling fluid in chemical processes, saving around 6.3 million m^3 of water per year, or 10% of the total consumption at the complex. Recycled water in Camaçari is cheaper and also less energy-intensive: each cubic meter of recycled water saves 1 kWh of electrical energy when compared to fresh water (Lima, Eduardo Pedroza da Cunha, 2018). The success of those two projects paved the way to a third one, in 2013, at the Duque de Caxias complex. This time it consisted of treatment and reuse of process water in cooling towers, saving 40,000 m^3 of fresh water per year.

Differently than in the cases of Camaçari or Duque de Caxias, Aquapolo has to deal with municipal wastewater from a densely populated area: its source is the sewage system from 1.4 million people. Municipal wastewater is rich in organic matter. It contains nitrogen, coming mainly from degraded proteins, which later on converts to ammonia/ammonium thanks to microorganisms, creating an additional oxygen demand. Ammonia/ammonium is then converted to nitrite and nitrate (biological nitrification process) causing further oxygen demand. The organic matter from municipal wastewater also releases phosphorus, which becomes ortho-phosphate. In addition to organic matter, soaps and detergents can also be a source of nitrogen and phosphorous. If ammonia and phosphorous are not properly removed from wastewater, when discharged into lakes, rivers, or estuaries, it will cause the addition of nutrients in the form of nitrates and phosphates and potentially degrade the aquatic environments.

The presence of oil and fats is another problem of this domestic wastewater. That contaminant increases the organic load of the effluent (BOD – Biological Oxygen Demand) and has low solubility, which makes it difficult for its biodegradation. It also has low density, so it can form a film that prevents the transference of oxygen from the air to the water. The average concentration of oil and fats in this region's raw wastewater is 69 mg/L, which is lowered to 6 mg/L after treatment in the conventional wastewater plant (Rodrigues, 2007). The presence of emergent contaminants is another rising concern in municipal wastewaters. These pollutants consist of chemical compounds from pharmaceuticals, personal care products, microplastics, and other substances that may end in the households' sewage (Karapanagioti and Kalavrouziotis, 2019). Some of them can undergo the wastewater treatment processes without degradation and pose a risk (e.g., acting as hormonal disruptors) when discharged into aquatic environments. Further research is still necessary to better assess their effects and the eventual improvements in wastewater treatment processes to deal with them. However, the potential presence of emerging contaminants is not a key factor when considering the use of recycled wastewater for industrial purposes such as cooling towers, washing equipment, and other non-drinkable purposes, as is the case of this petrochemical complex in São Paulo. Following, Fig. 1 shows an aerial view of Aquapolo facilities, which are located next to the municipal wastewater treatment plant.

The municipal wastewater arrives at a conventional treatment plant where it undergoes an activated sludge/deep shaft treatment that removes 88% of the BOD (Ministry of Cities of Brazil, 2017). The process in the conventional treatment plant consists of:

1. Screening to retain solids larger than 1 cm
2. Wastewater passes through a 4.5 m high sand filter
3. Primary settling tank

Fig. 1 The activated sludge treatment plant that already existed in 2012 (in *yellow (dark gray in print version)*) and the new Aquapolo units (in *blue (light gray in print version)*) (Asseburg, 2018).

4. Aeration tanks that promotes the microbial growth in the wastewater, and
5. Secondary settling tank (final clarifier)

The sludge that is generated in the decanters is sent to a filter press where water is recovered, and the resulting solid cake is sent to a landfill. Fig. 2 shows the flowchart of the processes performed at this conventional treatment plant.

The flow of effluent that leaves the activated sludge treatment plant is 2000 L/s. Its low BOD makes it complying with Brazil's quality parameters so it is suitable to be discharged in the nearest river, the Ribeirão dos Meninos. At this point, 650 L/s are diverted to the Aquapolo reclamation plant, and the rest is discharged at the river. There are sensors installed in this pipe to monitor the quality of the treated wastewater that is diverted to Aquapolo, so the reclamation plant can adjust the processes to better deal with variations in the characteristics of its influent. Real-time measures from PH, conductivity, and turbidity sensors are transmitted online to the plant's supervisory control and data acquisition system.

The activated sludge treatment plant was projected for the removal of carbon and nutrients, but not for effective removal of ammonia and phosphorous. To further polish the quality of the recycled wastewater to make it suitable for the petrochemical complex, the need to remove ammonia arose. To cope with that need, Aquapolo's process combined a biological treatment with ultrafiltration membranes in a Tertiary Membrane Bioreactor (TMBR), the first one in Brazil at the time. This tertiary treatment avoids discharging 0.6 tons of ammonia and 31.4 tons of phosphorous to the Ribeirão dos Meninos river, per year (Ambiental, 2013). Fig. 3 shows the flowchart of the processes performed at Aquapolo.

1. The first stage at Aquapolo is microfiltration. Wastewater is pumped throughout disk filters to remove particles larger than 400 μm. The aim is to remove solids that could harm future stages of membrane filtering (ultrafiltration and RO membranes).
2. The second stage consists of the Tertiary Membrane Biological Reactor (TMBR), which is shown in Fig. 4. The TMBR unit performs the biological treatment, with bacteria degrading organic matter in both aerobic and anaerobic zones. Besides organic matter, nitrogen and phosphorous are also removed. The biological reactor consists of a series of tanks that altogether have a volume of 5.350 m^3. This digestion of organic matter is followed by ultrafiltration throughout membranes that retain solids and bacteria larger than 0.05 μm. The membranes are encapsulated in nine tanks with 256 m^3 each.

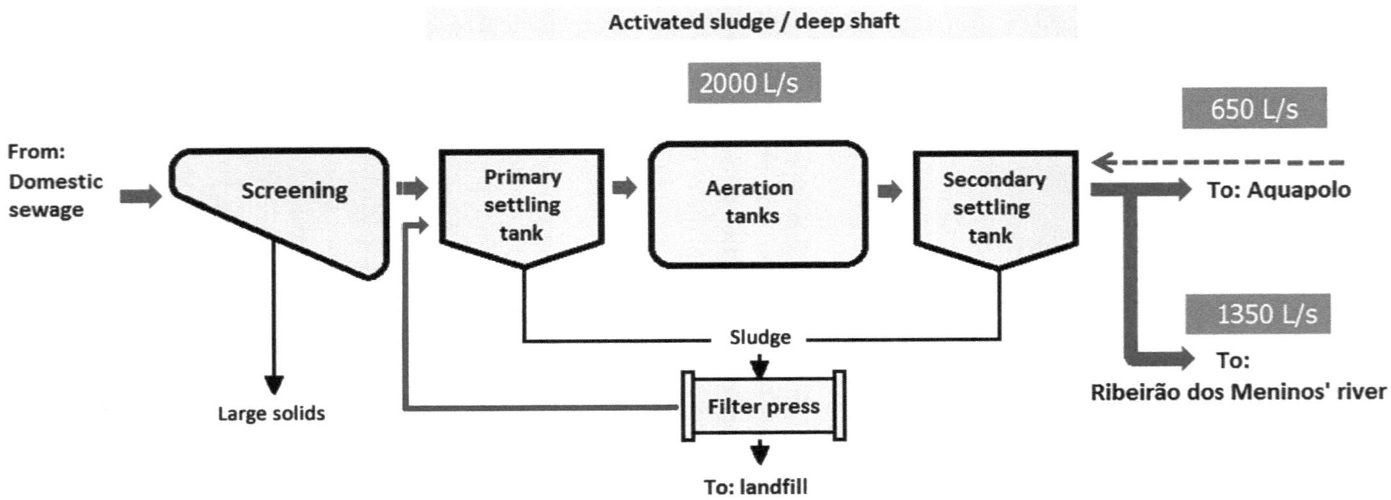

Fig. 2 Flowchart of the conventional treatment plant located previous to Aquapolo.

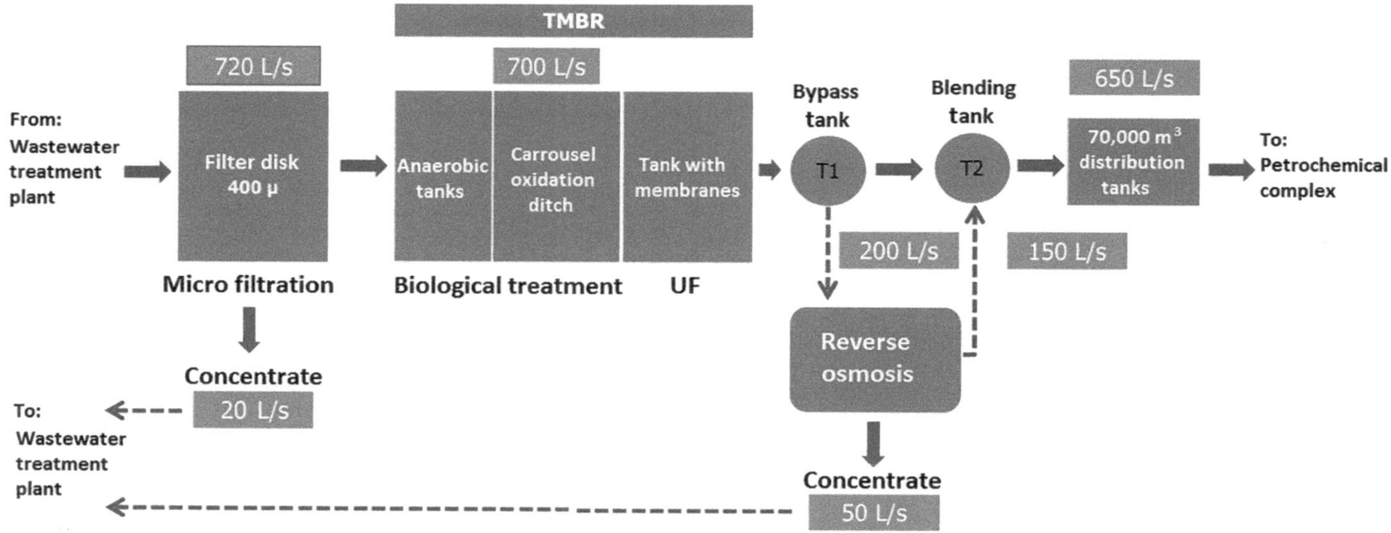

Fig. 3 Flowchart of Aquapolo. Current production is 650 L/s but the total capacity is 1000 L/s.

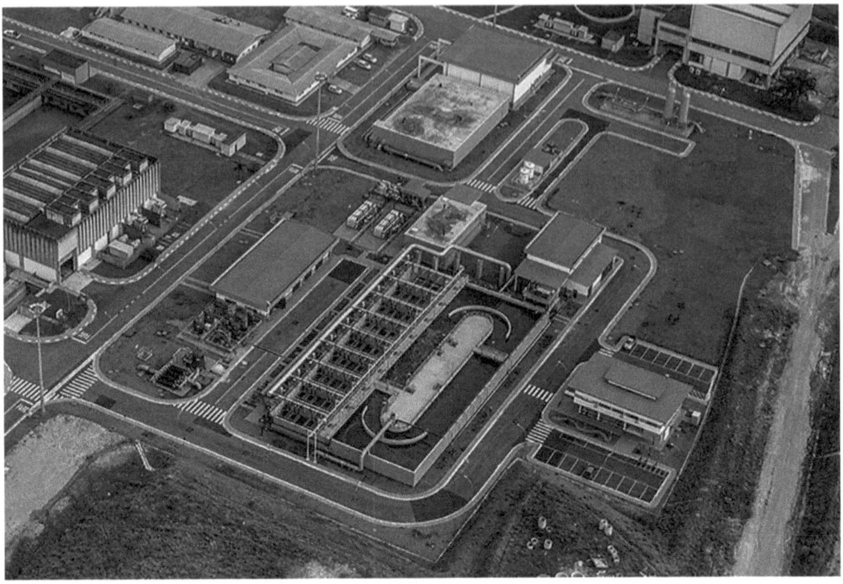

Fig. 4 The tertiary membrane biological reactor (TMBR) at Aquapolo.

3. The reverse osmosis (RO) system is actuated when the conductivity of the water exceeds the specification value of 50 μS/cm. If that is the case, the water stream is by-passed to two RO membrane systems that produce 150 L/s of permeate.
4. The reclamation plant is equipped with an automated disinfection system that doses chlorine dioxide (ClO_2) in key points of the process. When arriving at the petrochemical complex, the recycled water must have a residual concentration of chlorine dioxide higher than 0.2 mg/L.

Another differential of Aquapolo is the high level of automation and its supervisory control and data acquisition system. Online sensors monitor key quality parameters before, during, and after the plant. Programmable Logic Controllers (PLCs) linked to variable frequency devices control the rotational speed of the pumps to keep the process' flow adjusted to an optimal value. For example, if the organic load of the wastewater is too high and the retention time in the biological reactor needs to be incremented, the rotation of the pumps is automatically reduced to low the process' flow. That control strategy guarantees adequate values of BOD and COD. Sensors of conductivity control automated valves that will automatically divert the flow to the RO units when necessary, to keep the conductivity under the specified limit. The reclamation plant is equipped with 54 in-line analyzers for supervising and controlling the quality of the recycled water:

- 8 instruments for the analysis of total suspended solids,
- 7 for the analysis of turbidity,
- 11 for the analysis of PH,
- 15 for the analysis of conductivity,
- 7 for the analysis of chlorine dioxide (ClO_2),
- 1 for the analysis of oxidation–reduction potential (ORP),
- 2 for the analysis of dissolved oxygen (DO),
- 2 for the analysis of ammoniacal nitrogen, and
- 1 for the analysis of ferrous ions Fe(II) and Fe(III).

Besides the in-line analyzers, laboratorial analysis is periodically performed, regarding the following parameters: aluminum, copper, iron, manganese, BOD, COD, phenol index, surfactants, oils and fats, hardness, silica, and sulfides. For each of its clients, Aquapolo also makes available in-line measures of turbidity, pH, conductivity, chlorine dioxide, flow and pressure of the recycled water throughout a PLC installed at the delivery points in the petrochemical complex. The clients at the complex established the standards of water quality that wanted to be achieved. Table 2 shows the values of the quality parameters for the supplied recycled water (contract value) together with its average value as measured during 12 consecutive months. Note also that in the first column are specified the values of the quality parameters referent to the influent that enters the Aquapolo reclamation plant.

As shown in Table 2, the treatment process at Aquapolo has succeeded so far in meeting the client's requirements. The tertiary treatment, combining advanced biological and membrane processes, polishes the secondary effluent by removing ammonia and phosphate as well as inorganic pollutants like silica. The latter is a major issue when considering recycled water for industrial purposes, as silica forms scale deposits that can be hard, glassy, adherent, and difficult to remove from the internal heating surfaces of boiler units, for example. Silica deposits are typically formed when steam is cooled by expansion. Regarding cooling towers, if the tower fill has scaling, that deposit can reduce the tower's performance by minimizing the amount of air the tower fan can pull through to efficiently cool the bulk water. That would result in a reduction in heat transfer capacity affecting the process' performance, an increase in power consumption, and finally an increase in water consumption (more make-up water needed). The quality of recycled water that is achieved at Aquapolo allows for its application in cooling towers as well

Table 2 Specifications of the industrial water by the companies at the petrochemical complex supplied by Aquapolo (da Silva, 2015; Asseburg, 2018).

Parameter	Influent: treated wastewater	Effluent: recycled water		Unit
		Contract value	Average value	
Tenso-active compounds (surfactants)	5.1	1.0	0.16	mg/L
Aluminum	0.2	0.2	0.02	mg/L
Copper	0.1	0.1	0.008	mg/L
Conductivity	650	500	480	μS/cm
BOD – Biological Oxygen Demand	30	10	–	mg/L
COD – Chemical Oxygen Demand	100	20	8.4	mg/L
Total hardness	100	100	57	mg/L
Phenol index	0.130	0.134	0.040	mg/L
Total iron	1.5	0.3	0.16	mg/L
Total phosphorous	5	0.5	0.49	mg/L
Manganese	0.2	0.2	0.01	mg/L
Ammoniacal nitrogen	20	1	0.16	mg/L
Oils and Fats	10	<5	<5	mg/L
pH	5–9	6.5–7.5	6.78	–
Silica	20	20	5.75	mg/L
Total suspended solids	40	<5	<5	mg/L
Sulfide	0.9	0.1	0.02	mg/L
Turbidity	15	1	0.12	NTU
Residual concentration of chlorine dioxide	5.1	>0.2	0.26	mg/L

as for steam production, replacing fresh water. Fig. 5 shows an aerial view of some cooling towers at the petrochemical complex.

Aquapolo's recycled water has better quality than the previously used fresh water in terms of suspended and solved minerals (ions, hardness, silica content, and total suspended solids), which added additional advantages to the already lower price per cubic meter of process water. The first one was the increase in the number of cycles of concentration, that is, the number of times that cooling water can be used or pumped around (recirculated) before the water has to be blown down or bleed off. In cooling towers, part of the higher concentrated cooling water is drained from the system and replaced by feed water, which holds far less minerals. The feed water (makeup water)

Fig. 5 Cooling towers in the petrochemical complex, using recycled water supplied by Aquapolo.

dilutes the left cooling water and so brings down its overall mineral concentration, preventing the formation of deposits. With the use of recycled water, the number of cycles of concentration was increased on average between 1 and 2 cycles, resulting in less water consumption as less makeup water was needed to replace the water evaporated and bleed off.

As another additional advantage, some petrochemical units also experienced economic savings as lees chemical products – corrosion and scale inhibitors – were needed to treat the cooling water that is recirculated. One of the units at Braskem (UNIB 3 ABC) reported savings of 0.9 million USD per year in treatment products for cooling water (Ambiental, 2013). The better quality of the recycled water also impacted positively the maintenance of heat transfers, reducing the need for cleaning and for part replacements. Companies at the complex estimated another 0.9 million USD per year in savings due to that cause (Ambiental, 2013).

After almost 8 years of successful operation, the Aquapolo reclamation plant currently delivers 650 L/s of recycled water, from a total capacity of 1000 L/s. Therefore, there is spare capacity to increase the portfolio of new customers. The initiative was born to provide a steady and cheaper supply of high-quality water to the petrochemical complex. It was soon

Fig. 6 Water tanker truck for the supply of recycled water.

expanded to Paranapanema (2015) and Bridgestone Americas, Inc (2018) units, located outside the complex, throughout a second distribution network derived from Aquapolo's main pipe. Since then, no other large industrial consumer has been connected to the distribution network. Besides cooling towers and steam generation Aquapolo's recycled water can also be used in irrigation, cleaning of airplanes, vehicles, and streets, or civil construction, among other non-drinkable uses. The supply of recycled water through tanker trucks is an alternative for companies in the region that are not connected to Aquapolo's pipe network. So far some real estate developments have used that alternative, contracting volumes of recycled water up to 200 m^3/month and using it for street cleaning and vehicle washing during the construction phase of their buildings. Following, Fig. 6 shows one of Aquapolo's water tanker trucks.

4 Outcomes of the Aquapolo project

Aquapolo sets a milestone in Brazil with regard to water reuse and back in 2012, and since then, it still remains as the largest reclamation plant in the country. Other remarkable projects that produce recycled water from municipal wastewater are Capivari II, in the city of Campinas, and Prolagos, in Armação dos Búzios. Capivari II treats an average flow of 182 L/s of municipal wastewater and also uses a tertiary treatment through a membrane bioreactor, capable of nitrogen and phosphorus removal (Pagotto et al., 2014). It was fostered by a context of local water scarcity due to rapid urban expansion as well as by the existence of municipal laws that prohibit the use of drinking water for street washing, private car washing, and garden irrigation (Pagotto et al., 2014; Stepping, 2016). At the same time, Prolagos is much smaller (0.77 L/s) and uses reverse osmoses to polish treated wastewater for irrigating a golf course.

It is worth mentioning two important reclamation projects that, if implemented, could minimize the use of fresh water for industrial purposes in Rio de Janeiro, the country's second main city. One of them considered using 650 L/s of the rejected water outlet from the filters of Guandu Water Treatment Plant – the largest water treatment plant in the world – and treat that stream with flocculants, clarifiers and membranes to achieve the quality parameters of industrial water. A second project was very similar to Aquapolo as considered upgrading an existing wastewater plant with tertiary treatment to produce up to 1500 L/s of recycled water. A pipe would supply the recycled water from the upgraded Alegria Reclamation Plant to the clients located in a gigantic petrochemical complex under construction. A change in the economic situation ended up canceling the construction of the complex, so both projects did not come out. However, those two reclamation projects could be retaken in case of further deterioration of Rio de Janeiro's water sources (Obraczka et al., 2019).

As the leading project in the country, Aquapolo's experience also contributed to the discussion regarding the normative frame of the use of recycled water in Brazil, which is still in progress (Consortium CH2M Hill BV, 2016). So far, there is an absence of national legislation for reuse that goes beyond some municipal laws that only address where non-drinkable water can be used. A recent municipal law in São Paulo, approved in February of 2020, expands the allowed non-drinkable uses and defined two categories for recycled water, namely "type A" and "type B", depending on its quality parameters. The Municipal Chamber that developed this law counted with the participation of specialist from BRK, the company that operates Aquapolo (São Paulo State Government, 2020).

Another question raised by Aquapolo is the taxation of recycled water through the ICMS tribute (currently at 18% in São Paulo). The ICMS is a tax on sales and services that applies to the movement of goods, transportation, communication services and other general supplying of goods. It is, thus, a value-added tax (VAT). The Brazilian States stopped applying that tax to the drinkable water in the early 2000s, after a public discussion about whether water is a commodity or a basic good that is vital for human well-being. The consortium that operates Aquapolo, together with other

Brazilian water utilities, want to expand that tax exemption to recycled water for industrial purposes. It is worth mentioning that another Brazilian States that suffers from water scarcity, Ceará, is already discussing laws that exempt from ICMS tribute not only the recycled water, but "the acquisition of equipment for the building, operation and upgrading of reuse water facilities" (proposal nº 26/2016, Law nº 12.670).

Another feature of Aquapolo is that it established a precedent in the country by associating the city's water utility (Sabesp) – a public company – with a private company that invested and built the upgrades in the existing wastewater facility (BRK). That joint venture proved to be economically viable and that success could foster similar initiatives throughout the country. Among the factors that contributed to the success of the enterprise, some analysts have mentioned the use of an EPC (Engineering, Procurement and Construction) contract between BRK and Sabesp, together with the use of a take-or-pay contract for the supply to the main clients (Consortium CH2M Hill BV, 2016).

In addition, the project yielded considerable know-how to Sabesp and BRK. Both companies are engaged in the use of new technologies. Sabesp is responsible for supplying water to 28 million people and has to deal with a context of very limited water resources. The company has heavily invested in technologies for the reduction of urban water loss (Sánchez et al., 2020) and also prioritized projects for the reuse of municipal wastewater for non-potable purposes. Some of those reclamation projects were close to become reality during the 2013–2015 drought, but were finally turned down in favor of other alternatives (Stepping, 2016). Sabesp has developed water recycling projects since 2013 in four other wastewater treatment plants: Barueri, São Miguel Paulista, Jesus Neto, and Parque Novo Mundo. The latter supplies recycled water to textile industry and a pulp mill. Altogether, the total production of recycled water in Sabesp was 785 L/s in 2018, from which 650 L/s were due to Aquapolo. The utility has still a wide margin to further develop this activity, as currently only 0.43% of its treated wastewater is sold as recycled water (Sabesp, 2019).

At the same time, BRK currently is the largest private water and sanitation utility in Brazil. It has pioneered the introduction of the Nereda® technology for wastewater treatment in Brazil and is now operating the largest of this type of plant in the country (Khan et al., 2015). After Aquapolo, BRK started using recycled water from another reclamation plant (Mauá) to wash streets and sewer networks. BRK was acquired in 2019 by GS Inima Brasil.

Overall, Aquapolo stands as a technological milestone for Brazil's sanitation, as introduced the technology of membrane/bioreactor tertiary treatment and brought the installation of the first large-scale recycled water network in the country.

5 How the Aquapolo project helped fostering water efficiency and internal reuse projects among the participating companies

For the industry that buys the recycled water, the project succeeded in providing a continuous supply of high-quality input at an affordable price. That increase in water security showed to be crucial during the 2013–2015 drought.

But moreover, the steady reduction in expenses with water input encouraged the companies to further advance their own initiatives, such as internal effluent reuse projects. There has been an increase in the number of initiatives aiming for more efficient use of water since Aquapolo started its operations in 2013. Since 2017, it is organized every year a "Seminar of Sustainable Practices" where the companies that operate in the complex are invited to share the outcomes of these initiatives (COFIPABC, 2019). This event creates a framework for discussing the best practices and for collaborating in the management of water and effluent streams. All these initiatives resulted for the chemical plants in a continuous decrease of the water indicator index (m^3/ton), which consists of the cubic meters of water input that is needed to produce each ton of product output (Gallo, 2019; COFIPABC, 2019; Brancaglião, 2019). The industrial complex even saw the beginning of agreements between companies so the effluent of one industry, after proper treatment, is conducted to the next one that uses it as industrial water. After a process of continuous evolution between the years 2013 and 2020, the high level of effluent reuse allowed some industrial facilities to finally achieve ZLD, or zero effluents discharged to the environment. Table 3 lists the companies supplied by Aquapolo Reclamation Plant.

From the companies supplied by Aquapolo, only Bridgestone and Paranapanema are located outside the petrochemical complex. Those two companies have also shown its engagement in developing water efficiency and internal reuse projects. Paranaparema's copper smelting plant has its own water reclamation unit. It collects and treats stormwater, together with process water and wastewater from the administrative offices, supplying 81% of the total water consumption. The other 19% mostly comes from the

Table 3 Clients supplied by Aquapolo (da Silva, 2015).

Client	Company's activity	Flow of supplied water (L/s)
Braskem	Basic chemical compounds, polymers, and thermoplastic resins	537
Oxiteno	Ethylene oxide derivatives	44
Bridgestone	Tires	26
Cabot	Carbon blacks	18
Air Liquide/ Oxicap	Industrial gases	10
White Martins	Industrial gases	9
Paranapanema	Copper co-products	6
Total		650

recycled water bought from Aquapolo (Paranaparema, 2019). At the same time, Bridgestone's industrial plant claims to treat 100% of its effluent and reuse it as process water (Bridgestone Americas, Inc, 2018). The largest consumption of water in this tire factory is due to steam generation, which comes exclusively from recycled water, either produced internally or bought from Aquapolo.

Braskem outstands as the main consumer of recycled water and was since the beginning one of the main stakeholders of Aquapolo. Its commitment to reduce fresh water withdrawals yielded its inclusion in the CDP's "Water Security A List" (CDP Global, 2019). Besides developing internal reuse projects, Braskem's water strategy has also relied on partnering with outsourced environmental management companies. That has been the case with Cetrel, in Camaçari's Petrochemical Complex, and with BRK – the company that operates Aquapolo – in São Paulo. Both Braskem and BRK are active members of "Aquasfera", a digital hub that assesses water security in the Brazilian industry. That digital hub is an initiative created by the Brazilian Business Council for Sustainable Development (CEBDS) and arises from the signed commitments during the 8th World Water Council of Brasilia in 2018 (CEBDS, 2018).

The second largest consumer of Aquapolo's water is Oxiteno's industrial plant. It has a water indicator index of 4.31 m^3/ton of product but, since 97% of that consumption comes from recycled water, the freshwater indicator is much lesser: only 0.13 m^3 of fresh water per ton of product (Oxiteno, 2019). The average rate of recycled water use for all Oxiteno plants worldwide is 15.2%, so this one at São Paulo's complex outstands as one with the lowest impact on freshwater resources of the company (Oxiteno, 2019).

The freshwater indicator expresses the actual volume of fresh water that needs to be withdrawal to produce one ton of chemical product. For the companies that operate in the complex (Braskem, Oxiteno, Cabot, Air Liquide/Oxicap and White Martins), the average freshwater indicator was 0.7 m^3/ton of product (until 2017) and 0.6 m^3/ton of product from 2018 onwards (COFIPABC, 2019).

Data shows that the withdrawal of fresh water, whilst being well below the average for the chemical industry, is still significant. In 2018, the petrochemical complex bought 1.2 million m^3 of fresh water and 9.7 million m^3 of recycled water. Therefore, Aquapolo was responsible for 89% of the total water supply of the complex while the other 11% was fresh water coming from groundwater wells and the city mains (COFIPABC, 2019). Although the reduction of freshwater withdrawal by the facilities has been impressive since 2013, some cases indicate that further improvements could be achieved in that direction. In fact, one industrial facility in the petrochemical complex has evolved to a point of reaching ZLD. Hereafter it is presented the case of Cabot's chemical plant, which managed to achieve ZLD after a major reuse project in 2016 that rearranged the streams of water and effluents in the facility. Since 2016, this plant recovers 100% of its liquid effluent and reuses it in its industrial process.

The main product of this facility is carbon black, which can be used in a wide range of compounds from reinforced elastomers in tires, to pigments in inks, coats, and plastics. This facility has three production lines that yield a total capacity of 140,400 tons of carbon black per year. Carbon black is produced by the controlled vapor phase pyrolysis of hydrocarbons in a reactor. The resulting gaseous stream is filtered to retain the carbon black solids, which then undergo milling, pelleting, and drying. After removing the solids from the process gas, most of the waste heat contained in the exhaust gas is recovered with the use of a turbine, heat exchangers, and a boiler (Vallerio, 2019). These utilities allow the facility to increment its energy efficiency but require a significant amount of water. Water is needed in the process as a cooling fluid and also for washing the equipment. Water input comes from various sources while water output can leave the facility in three ways:

(I) evaporated to the atmosphere in the cooling equipment,
(II) as a liquid effluent stream (wastewater), or.
(III) incorporated in the product.

To understand the water balance of this industrial facility, it is presented a flowchart that describes the flow of water in and out of it. Thus, Fig. 7 illustrates the flowchart of water as in 2015, before the internal reuse project.

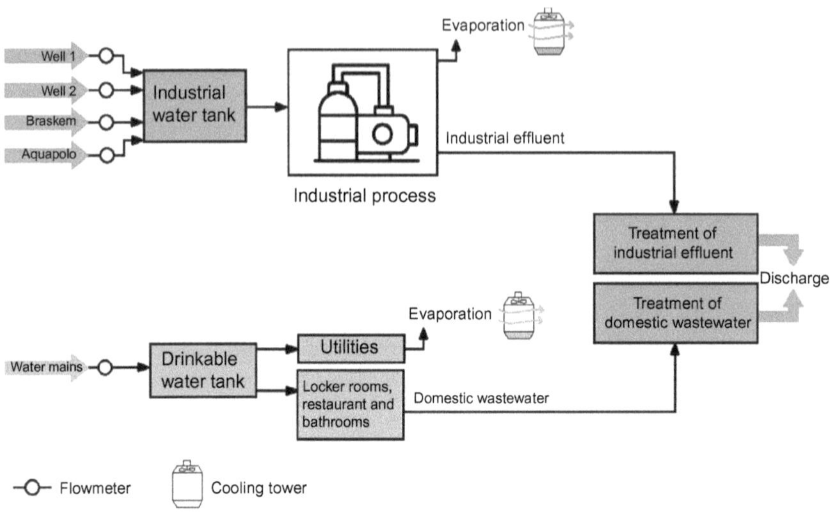

Fig. 7 Water balance at Cabot's industrial facility before the internal reuse project (2015).

As shown in Fig. 7, there were two inputs of recycled water entering the facility: one coming from Aquapolo and the other from another facility in the complex (Aquapolo Ambiental, n.d.). Recycled water was mixed with groundwater from two wells when necessary and then used as industrial water. The bottom part of the figure shows a freshwater input coming from the city mains. It supplied water for drinkable purposes in the restaurant and administrative buildings as well as for non-drinkable purposes in some cooling and power utilities. Every water input was measured by a flow meter. Regarding the water outputs, evaporation in cooling systems represented 65–80% of the overall water consumption. The water output incorporated into the product was only 1%. Therefore, the rest of water consumption (19–34%) left the facility as domestic and industrial wastewater. "Domestic wastewater", generated by around 150 people, was much lesser in volume than the industrial wastewater. The greatest fraction of the liquid effluent was "industrial wastewater" whose main contaminant was pure amorphous carbon.

The industrial effluent came from the water used to c1ean or purge tanks and other equipment used in batch-type operations. Upon contact with these surfaces, the water carries particles of carbon black. This substance is not soluble, and when released into water, it is expected to be present as suspended particulate matter that will eventually settle into bottom sediments (Government of Canada, 2013). As the substance has a specific gravity of 1.7–1.9 (water = 1), gravity settling is effective and is the most common technique employed to remove it from wastewater (ICBA, 2016). Flocculating agents, such as ferric or aluminum sulfate, or synthetic polymers, can also be used to enhance settling. Therefore, a proper treatment for an industrial effluent containing carbon black is the use of settling tanks and clarifiers to reduce turbidity and remove the settled particles. The overflow of the clarifiers can then be further polished with the use of membrane filtering. A microfiltration membrane can retain suspended solids or colloidal particles between 0.1 and 10 μm in diameter, while the ultrafiltration process commonly retains particles between 0.001 and 0.1 μm. The primary carbon black particle, also known as the nodule, is approximately 10–500 nm in diameter (Wang et al., 2003). However, during the production process, the carbon black nodules coalesce to form aggregates that are about 80–810 nm in size and later form agglomerates of hundreds to thousands of adhering aggregates of 1–100 μm in diameter (OECD, 2006; ICBA, 2016). Carbon black also has a tendency to further agglomerate to avoid contact with water (IARC, 1996). As a result, the average particle size that is present in the effluent is typically around 100 μm so most of this pollutant can be effectively removed through ultrafiltration (Lohwacharin et al., 2009; Qin et al., 2020).

The reuse project that was implemented in 2016 introduced major rearrangements on the streams of both domestic and industrial effluents. Since the project was implemented, domestic and industrial wastewaters are treated separately and then mixed to produce "industrial water" that must comply with the internal quality standards of the process, which are more rigid than the national norms for wastewater discharged to surface waters (CONAMA, 2011). Parameters, such as total suspended solids (TSS) and total dissolved solids (TDS), are given special attention.

Since 2016, stormwater is also collected and mixed with the industrial effluent containing particles of carbon black. That mixture is later treated through settling tanks. Rainwater arose as a new water source for the chemical plant, but before introducing it in the treatment system, it is important to take some precautions. Stormwater is collected from the roofs and grounds around the facility and conducted to two large ponds. As the initial runoff of stormwater is retained in the ponds, most of the largest particles and organic matter are removed there (Sánchez et al., 2015a, b). A much cleaner stream of stormwater is then conducted to the settling tanks where it mixes with the industrial effluent. In addition, the ponds prevent the flow of stormwater from introducing turbulence into the settling tanks. After the removal of suspended solids, this mixed stream goes to an "internal reuse water tank" that was also built in 2016. At the same time, the domestic wastewater undergoes a biological process and then disinfection, before being conducted to the same internal reuse water tank.

One of the main changes after 2016 is that no more drinkable water is used in the cooling utilities. Since then, all the cooling water is supplied by either the previously existing "industrial water tank" or the new "internal reuse water tank".

As a result of these upgrades, 100% of both domestic and industrial effluents in the plant are reused. Cabot's industrial facility became a ZLD plant, which means that no liquid effluents are discharged to the environment. All the input water is either lost by evaporation or incorporated into the product. Fig. 8 illustrates the flowchart of water after the internal reuse project.

In 2015, the water sources at Cabot's facility were distributed as follows: 79.4% was recycled water (supplied exclusively by Aquapolo reclamation plant), 6.1% was drinkable water coming from the city mains, and 14.5% came from two wells operated by the company.

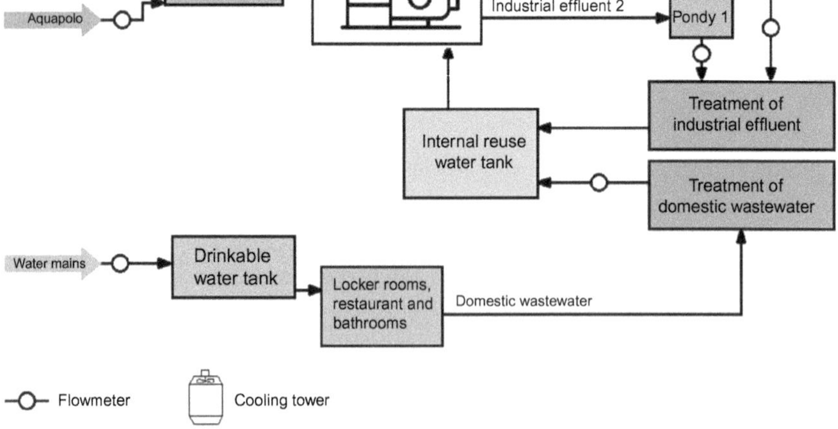

Fig. 8 Water balance at Cabot's industrial facility after the internal reuse project (2016).

The internal reuse project brought a major change in the facility's water sources. In 2018, 80% of the total consumption was recycled water supplied by Aquapolo, 16.4% was recycled water from internal reuse, 3.2% was drinkable water coming from the city mains, and only 0.4% came from the wells. That means that 96.4% of the overall water sources at this industrial facility comes from the recycling of domestic wastewater, stormwater, and process effluents. The need for drinkable water from the city mains and the water extracted from the wells were set to the minimum.

The low impact on water resources achieved by Cabot's ZLD facility in São Paulo highlights it among all the chemical plants that this company has worldwide. Globally, 95% of the water used by Cabot Corporation in 2018 was withdrawn from surface sources or purchased from the mains, another 4% was withdrawn from groundwater, and only 2% came from treated wastewater from households (Cabot Corporation, 2019).

The transition of Cabot's facility in São Paulo toward a ZLD plant is reflected in the evolution of its water index indicators, as shown in Table 4.

The internal reuse project developed in 2016 allowed to reduce the consumption of water from all sources, but particularly from the city mains and groundwater.

Data shown in Table 4 indicate that the upgrades in the facility resulted in less freshwater withdrawal and a larger share of recycled water among the overall consumption. But, moreover, data also show that the chemical plant became more efficient in the use of water. The "Total water index" is an indicator related to how efficient in the use of water the process of a company is, given the water footprint of its products. Internal reuse projects can have a large impact on improving that index, as is the case: the index dropped 16.3% after the start-up of the reuse project in 2016. That index

Table 4 – Evolution of the water use index at Cabot's facility between years 2015–2018, separated by each water source (Gallo, 2019).

	2015	2016	2017	2018	Variation 2015–2018
Total water index (m^3/ton of product)	5.40	4.52	4.27	4.45	−17.6%
Recycled water (m^3/ton of product)	4.29	4.29	3.97	4.11	−4.2%
Drinkable water from the city mains (m^3/ton of product)	0.33	0.22	0.24	0.14	−57.6%
Groundwater extracted from the well (m^3/ton of product)	0.78	0.01	0.06	0.20	−74.3%

can also be reduced through improvements in the efficiency of the cooling equipment, leakage detection programs, better process control, or any other measure that optimizes the use of water in the industrial process. From 2016 onward, once the plant had already become ZLD, it managed to further reduce the total water index 1.5% on average.

6 Conclusions

- Typically, municipal wastewaters undergo a series of physical and biological treatments: settling, sand filtering, activated-sludge, aeration. The treated wastewater can then be discharged into water bodies or, instead, it can undergo a further process called tertiary treatment. It combines advanced biological and membrane processes and removes much of ammonia and phosphate as well as inorganic pollutants like silica.
- The reclamation of municipal wastewaters is a valid source of water for non-drinkable purposes in cities, agricultural fields, and industrial facilities.
- The use of recycled water adds a new business line to the portfolio of water utilities, while allows the industry to obtain a stable supply at a lower price. From a public perspective, reclamation of municipal wastewaters reduces the water stress of cities and should be fostered by municipal and national laws.
- To develop a successful water reclamation scheme, it is key to identify and partner with potential consumers that may use the recycled water.
- The chemical industry is a great potential consumer of recycled water. Most of its water consumption comes from three non-drinkable uses: washing equipment, steam production, and process cooling. These two latter uses require that the recycled water achieves low content of suspended and solved minerals, to avoid scaling and silica deposits. Constant monitoring of the quality parameters of the recycled water is a keystone to successfully supply industrial uses.
- Aquapolo in São Paulo is a successful case of cooperation between three stakeholders: (I) the buyers of recycled water at a petrochemical complex, (II) the water utility that operates the city's sewage network, and (III) a company that operates the facility that processes treated wastewater into recycled water, and supplies it to the industries at the complex.

- Aquapolo showed to be strategic for the industrial facilities during São Paulo's 2013–2015 drought. The stable supply of recycled water prevented them from economic losses due to halting their operations.
- In addition, data between 2013 and 2020 show that Aquapolo had a positive effect on fostering sustainable water use practices on the companies that supplies. Most of them developed their own internal reuse projects and managed to reduce their water intensity.
- The internal reuse projects implemented by these companies are based on the segregation of wastewaters. Domestic wastewater from the bathrooms and restaurants in the facility undergoes a biological process followed by disinfection. Industrial effluent is treated separately through settling tanks, flocculation, and membrane filtration when necessary. Many facilities at the complex are also collecting rainwater from roofs and open areas; after discarding the initial runoff, the stormwater is mixed with the industrial effluent and then treated. The two wastewater streams (domestic and industrial/rainwater) are treated separately and then mixed in "industrial water tanks" that supply non-drinkable water uses.
- Such an internal reuse project allowed a chemical plant in the complex to become a ZLD facility. It managed to treat and recirculate 100% of the liquid effluent generated in the plant, so nothing is discharged to the sewers. As a result, all the water input is either incorporated in the product or evaporated.
- The internal reuse of effluent within a facility reduces its need for water input, which is measured by the "total water index" (m^3 of water per ton of product).
- Conversely, the use of recycled municipal wastewater as a water input reduces the impact that the industry has on freshwater sources. The need for freshwater withdrawal is measured by the "fresh water index" (m^3 of fresh water per ton of product).

References

Ambiental, A., 2013. Inscrição do Projeto Aquapolo na 8ª Edição do Prêmio Fiesp de Conservação e Reuso de Água. 8° Edição do Prêmio de Conservação e Reúso de Água – FIESP/CIESP – Categoria Empresa de médio e grande porte. São Paulo.

Aquapolo Ambiental S.A. http://www.aquapolo.com.br/ Accessed 04/19/2020.

Asseburg, M.K., 2018. Case Aquapolo - Água de Reúso Industrial. In: IX Brazilian Congress of Environmental Management. São Paulo, November 26–29.

Brancaglião, L., 2019. Sustentabilidade na Oxiteno. Práticas Sustentáveis no Polo do Grande ABC, Mauá, Brazil, September 12.

Braskem Corp. https://www.braskem.com.br/eficiencia-hidrica-en. Accessed 03/24/2020.
Bridgestone Americas, Inc, 2018. Bridgestone reduces water consumption in its plants. Press Release. São Paulo, March 20.
Cabot Corporation, 2019. Sustainability Report 2018. Boston. Available at www.cabotcorp.com.
CDP Global, 2019. CDP Global Water Report 2018: Treading Water - Corporate Responses to Rising Water Challenges. CDP, London.
CEBDS, 2018. Brazilian Business Council for Sustainable Development. Brazilian Business Commitment to Water Security. 8th World Water Council. Brasilia, Brazil, March 18–23.
COFIPABC, 2019. Comitê de Fomento Industrial do Polo do Grande ABC. In: Indicadores de Segurança e Meio Ambiente. Práticas Sustentáveis no Polo do Grande ABC, Mauá, Brazil, September 11.
CONAMA, 2011. National Council for the Environment. Resolução CONAMA N° 430 DE 13/05/2011. Published in the "Diario Oficial da União" (Official Bulletin of the Brazilian Federation), on May 13, 2011. Brasilia, Brazil.
Consortium CH2M Hill BV, 2016. Document II (RP01A): Reuse Experiences. Brazilian Ministry of the Cities and IICA – Interamerican Institute of Cooperation for the Agriculture.
da Silva, F.G., 2015. Tecnologia e Sustentabilidade do Projeto Aquapolo. Ecogerma Congress, São Paulo, September 30.
Gallo, M., 2019. Sustentabilidade na Cabot. Práticas Sustentáveis no Polo do Grande ABC, Mauá, Brazil, September 12.
Government of NCT of Delhi, 2009. Chapter 13 - Water Supply and Sewerage. In: Economic Survey of Delhi, 2008–2009, pp. 151–163. Delhi.
Government of Canada, 2013. Environment Canada/Health Canada. Screening Assessment for the Challenge: Carbon Black. CAS RN 1333-86-4.
Hussain, C.M., 2019. Handbook of Environmental Materials Management, first ed. Springer International Publishing.
Hussain, C.M., 2020. The Handbook of Environmental Remediation: Classic and Modern Techniques. Royal Society of Chemistry, https://doi.org/10.1039/9781788016261.
Hussain, C.M., Keçili, R., 2019. Modern Environmental Analysis Techniques for Pollutants, first ed. Elsevier.
Hussain, C.M., Kharisov, B., 2017. Advanced Environmental Analysis-Applications of Nanomaterials. The Royal Society of Chemistry, https://doi.org/10.1039/9781782623625.
Hussain, C.M., Mishra, A.K., 2019. Nanotechnology in Environmental Science, 2 Volumes. John Wiley & Sons.
IARC - International Agency for Research on Cancer, 1996. Printing Processes and Printing Inks, Carbon Black and some Nitro Compounds. World Health Organization, Lyon (FR).
ICBA, 2016. Carbon Black User's Guide: Safety, Health, & Environmental Information. International Carbon Black Association, New Orleans.
International Council of Chemical Associations, 2015. Responsible Care Status Report.
Jacobsen, M., Webster, M., Vairavamoorthy, K., 2013. The Future of Water in African Cities - Why Waste Water? International Bank for Reconstruction and Development/The World Bank, Washington, DC.
Karapanagioti, H., Kalavrouziotis, I., 2019. Microplastics in Water and Wastewater. IWA Publishing, London.
Khan, A.A., Ahmad, M., Giesen, A., 2015. NEREDA®: an emerging technology for sewage treatment. Water Pract. Technol. 10 (4), 799–805. https://doi.org/10.2166/wpt.2015.098.

Lahnsteiner, J., Lempert, G., 2007. Water management in Windhoek. Namibia. Water Sci Technol 55 (1–2), 441–448. https://doi.org/10.2166/wst.2007.022.

Lazarova, V., Asano, T., Bahri, A., Anderson, J., 2013. Milestones in Water Reuse: The Best Success Stories. IWA Publishing, London. Available at: https://www.iwapublishing.com/books/9781780400075/milestones-water-reuse.

Lima, Eduardo Pedroza da Cunha, 2018. Water and Industry: Experiences and Challenges, first ed. Infinita Imagem, Brasília, Brazil.

Lohwacharin, J., Oguma, K., Takizawa, S., 2009. Ultrafiltration of natural organic matter and black carbon: factors influencing aggregation and membrane fouling. Water Res. 43 (12). https://doi.org/10.1016/j.watres.2009.04.018.

Ministry of Cities of Brazil, 2017. (absorbed in 2019 by the Ministry of Regional Development). Relatório de Esgotamento Sanitário Municipal – São Paulo. Atlas Esgotos, Despoluição de Bacias Hidrográficas.

Ministry of the Cities of Brazil, 2018. Sistema Nacional de Informações sobre Saneamiento. Brasilia.

Myers, J.E., 2014. Chevron San Ardo Facility Unit (SAFU) beneficial produced water reuse for irrigation. In: SPE International Conference on Health, Safety, and Environment, 17–19 March 2014, Long Beach, USA., https://doi.org/10.2118/168401-MS.

NRC - National Research Council, 2012. Water reuse: Potential for expanding the nation's water supply through reuse of municipal wastewater. In: Committee on the Assessment of Water Reuse as an Approach to Meeting Future Water Supply Needs. Water Science and Technology Board, Division on Earth and Life Studies, National Research Council, Washington, DC, https://doi.org/10.17226/13303.

Obraczka, M., Silva, D.R., Campos, A.S., Muricy, B., 2019. Wastewater reuse of secondary treatment as an alternative source of water supply in the city of Rio de Janeiro. Sistemas Gestão 14 (3), 291–309. https://doi.org/10.20985/1980-5160.2019.v14n3.1392.

OECD – Organisation for Economic Co-operation and Development, 2006. Carbon Black, CAS 1333-86-4; SIDS Initial Assessment Report for Carbon Black; CAS no. 1333-86-4. SIDS Initial Assessment Meeting 21, Washington (DC), 18–21 October 2005 (Includes SIDS Initial Assessment Profile, SIDS Initial Assessment Report, and IUCLID Data Set). United Nations Environment Programme, Geneva (CH). Available from: http://cs3-hq.oecd.org/scripts/hpv/.

Oliveira-Esquerre, K.P., Kiperstok, A., Kalid, R., Sales, E., Teixeira, L.S., Pires, V.M., 2009. Water and wastewater management in a petrochemical raw material industry. Comput.-Aided Chem. Eng. 27, 1047–1052. https://doi.org/10.1016/S1570-7946(09)70395-5.

Oxiteno, 2019. Sustainability Report 2018. São Paulo, Brazil.

Pagotto, R., Rossetto, R., Gasperi, R.L., Andrade, J.P., Trovati, J., Vallero, M.V., Okumura, A., Arntsen, B., 2014. SANASA Capivari II - the first full-scale municipal membrane bioreactor in Latin America. Water Sci. Technol. 70 (2), 272–278. https://doi.org/10.2166/wst.2014.204.

Paranaparema, 2019. Sustainability Report 2018. Paranaparema S.A. and Orbis Comunicação, Dias D'Ávila, Brazil.

Qin, H., Guo, W., Gao, P., Xiao, H., 2020. Spheroidization of low-cost alumina powders for the preparation of high-flux flat-sheet ceramic membranes. Ceram. Int. https://doi.org/10.1016/j.ceramint.2020.02.093.

Rodrigues, A.L.G., 2007. Effects of Oil and Fats in the Treatability of Wastewater and Diffuse Pollution. Sabesp.

Sabesp, 2019. Sustainability Report 2018. Sabesp Investors Relations, São Paulo, Brazil. Available at www.sabesp.com.br.

Saltgae Project, 2016. Horizon 2020, Call H2020-WATER-1b-2015. Grant Agreement n° 689785. http://saltgae.eu/.

SAMCO Tech, 2019. What Industrial Facilities Need to Know About Zero Liquid Discharge. Buffalo, NY.

Sánchez, A.S., Matos, A.P., 2018. Desalination concentrate management and valorization methods. In: Gude, V.G. (Ed.), Sustainable Desalination Handbook, first ed. Elsevier, Oxford, pp. 351–399, https://doi.org/10.1016/B978-0-12-809240-8.00009-5.

Sánchez, A.S., Nogueira, I.B.R., Kalid, R.A., et al., 2015a. Desalination 364, 96–107. https://doi.org/10.1016/j.desal.2015.01.034.

Sánchez, A.S., Cohim, E., de Araújo Kalid, R., 2015b. A review on physicochemical and microbiological contamination of roof-harvested rainwater in urban areas. Sustain. Water Qual. Ecol. 6, 119–137. https://doi.org/10.1016/j.swaqe.2015.04.002.

Sánchez, A.S., Oliveira-Esquerre, K.P., dos Reis Nogueira, I.B., de Jong, P., Filho, A.A., 2020. Water loss management through smart water systems. In: Patnaik, S., et al. (Eds.), Smart Village Technology, Modeling and Optimization in Science and Technologies 17. Springer Nature Switzerland AG, pp. 233–266, https://doi.org/10.1007/978-3-030-37794-6_12.

São Paulo State Government, 2020. Joint Resolution SES/SIMA - 1, from 13-2-2020. São Paulo, 130 (32), page 4. In: Published in the "Diario Oficial do Estado de São Paulo" (Official Bulletin of São Paulo State), February 14.

Seguí Amórtegui, L.A., Moeller-Chávez, G., De Andrés Mosquera, A., 2019. Mexico, the water stress: challenges and opportunities in wastewater treatment and reuse. In: Guerrero García Rojas, H. (Ed.), Water Policy in Mexico. Global Issues in Water Policy. vol 20. Springer, Cham, https://doi.org/10.1007/978-3-319-76115-2_4.

Siemens, J., Huschek, G., Siebe, C., Kaupenjohann, M., April 2008. Mexico City–Mezquital Valley. Water Res. 42 (8–9), 2124–2134. https://doi.org/10.1016/j.watres.2007.11.019.

Stepping, K.M.K., 2016. Urban sewage in Brazil: drivers of and obstacles to wastewater treatment and reuse. In: Discussion Paper 26/2016. DIE - Deutsches Institut für Entwicklungspolitik, Bonn.

United Nations Department of Economic and Social Affairs (UN DESA), 2018. World Urbanization Prospects: The 2018a Revision. UNDESA, New York.

Vallerio, R., 2019. Aproveitamento Energético de Gases Residuais. Práticas Sustentáveis no Polo do Grande ABC, Mauá, Brazil, September 11.

van der Merwe, B., 2009. Closing the urban water cycle - integrated approach towards water reuse in Windhoek, Namibia. In: Cisneros, B.J., Rose, J.B. (Eds.), Urban Water Security: Managing Risks. UNESCO-IHP, first ed. CRC Press, London, https://doi.org/10.1201/9780203881620.

Vigneswaran, S., 2009. Wastewater Recycling, Reuse, and Reclamation - Volumes I and II. UNESCO-Encyclopedia Life Support Systems (UNESCO-EOLSS). EOLSS Publications.

Wang, M.J., Gray, C.A., Reznek, S.A., Mahmud, K., Kutsovsky, Y., 2003. Carbon black. In: Kirk-Othmer Encyclopedia of Chemical Toxicology. John Wiley & Sons, New York (NY), pp. 761–803.

WEF – Water Environment Federation, 2018. The Water Reuse Roadmap. WEF, Alexandria.

Zero brine Project, 2017. Horizon 2020, Call H2020-WATER Grant Agreement n°730390. https://zerobrine.eu/.

CHAPTER TEN

Conclusions

Sunpreet Singh[a] and Chaudhery Mustansar Hussain[b]
[a]Mechanical Engineering, National University of Singapore, Singapore, Singapore
[b]Department of Chemistry & Environmental Science, New Jersey Institute of Technology, Newark, NJ, United States

Zero waste (ZW) is a set of principles focused on waste prevention that encourages the redesign of resource life cycles so that all products are reused, being an ultimate goal that no trash should be sent to landfills, incinerators, or the ocean. The present book provided invaluable insights into the broad horizons of the ZW Tools practiced so far in various commercial sectors. Indeed, the basic concepts of the ZW can be traced back to centuries ago; however, the dynamic innovations through which it has undergone has changed the modern practices with more emphasis on the real-life estimation of the tool's effectiveness. It has been observed that various R's, such as recycling, reuse, recover, redesign, reduce, and remanufacture, are the key pillars of any ZW strategies. ZW is not only promoting these R's but also the prevention and product designs that consider the entire product life cycle. However, there exist numerous questions regarding Real ZW. For instance:

- Is the Real ZW achievable?
- How to recognize the most efficient tool of Real ZW?
- Can landfilling be completely avoided?
- Do commercial manufacturing technologies efficient enough?
- What are the various threats to the small and medium scale enterprises adopting ZW tools?
- How to address the socio-economic, environmental, and health concerns?

To address some of the above-mentioned questions, researchers have developed various continuously evolving hierarchies for addressing the set goals of different commercial sectors. However, most of these end at dumping the unrecovered waste that increases environmental hazards. On a recent record, the much-appreciated consideration of Industry 4.0 has emerged as a vital tool to acknowledge the existing socio-economic and environmental concerns. Through Industry 4.0, it is possible to eliminate various intermediate steps involved in the production of a good or service; therefore, resources could

Concepts of Advanced Zero Waste Tools
https://doi.org/10.1016/B978-0-12-822183-9.00010-6

be saved significantly. Furthermore, in the developing world, there is a scarcity of government legislation and standard practices to deal with the waste. It has been addressed by the specialists that standardized procedures for dealing with waste should focus either on the opened-loop or closed-loop recycling processes so that waste could be recovered at a greater throughput. Despite this, the fundamental transformation of the societies should be aimed at ZW through closed-loop waste treatment systems in place. The nonrecyclable waste should be subjected to waste treatment plants, wherein the materials, in the form of heat or energy, could be recovered.

The local authorities and/or regional municipalities can play a supporting role in tackling the sustainability issues of the ZW tools, as these can provide required infrastructures, machinery, and waste management. Furthermore, routine campaigning for public awareness and youth education can be planned by the authorities and municipalities. Furthermore, media networks and celebrities could also be involved in such awareness campaigning. Moreover, consulting and training activities on sustainable waste management targeting ZW must be the top priority to support related initiatives. For instance, through sustainable product packaging systems, daily business practices to reduce waste and recycling expenditures. Further, the context of sustainable packaging can meet the future community and industry challenges.

On this line, many countries have already banned plastic-based packaging materials and adopted biodegradable packaging systems. There must be a fundamental shift in thinking of the modern societies so that prioritized set constraints of the industry and society can be focused on bringing rapid and meaningful changes. Closely related, the fashion industry has caused unbearable damage to the environment. Therefore, fashion brands should incorporate sustainable development into their business strategies and recycling schemes. On the part of consumers, various approaches have identified that shopping behaviors matter a lot in waste control. Responsible and sustainable consumer behavior is also another key pillar in waste prevention.

Overall, the ZW is both an inspiration and resource terminology to help industries and societies to collaborate on the adoption of sustainable practices which can reduce the waste and work for the circular economy using in-process material wastages.

Index

Note: Page numbers followed by *f* indicate figures and *t* indicate tables.

A

Acid hydrolysis, 98
Activated sludge treatment plant, 224*f*, 225
Additive manufacturing (AM), 16*f*
Adhesives/cosmetics/films, 164
Aerators, 165–166
Africa, zero waste in, 137
Agriculture, feather keratin in, 101
Air Liquide/Oxicap, 236–237, 236*t*
Alegria Reclamation Plant, 233
Alkali hydrolysis, 98
Alkaline hydrolysis, 160
Alkaline lipase, 155
Alkaline protease enzymes, 156
Alpha helix, 102
AM. *See* Additive manufacturing (AM)
Ammonium nitrate (AN), 56
Ammonium sulfate (AS), 56
Anaerobic codigestion, 104–105
Anaerobic digestion, 104–105, 109–110
Animal feed/chicken feed/fertilizer, 161–162
Anthroposphere, 187
Aquapolo wastewater reclamation plant, 220–221, 224*f*, 242
  advantages, 231
  Braskem and, 222
  Camacari unit, 222–223
  clients, 235, 236*t*
  cooling towers in petrochemical complex, 229–230, 231*f*
  data of, 221*t*
  flowchart of, 225–228, 227*f*
  history, 221
  in-line analyzers, 228–229
  laboratorial analysis, 229
  outcomes, 232–235
  recycled water
    quality parameters, 229, 230*t*
    taxation, 233–234
    water tanker truck, 231–232, 232*f*
  sludge treatment, 224–225
  Tertiary Membrane Bioreactor, 225, 228*f*
Aquasfera, 236
"Arava" project, 216
Asparagine, 98
Aspergillus tamarii, 155–156
Australia, ZW development in, 77
Autoclaved hatchery wastes, 104–105

B

Backyard poultry birds, 91–93
Bangladesh, 3R model in, 75–77
Basic chromium sulfate (BCS), 157
Basified salts, 157
BCS. *See* Basic chromium sulfate (BCS)
Beam house operation, 145–146
Beta-sheets, 95–97
"Beyond the zero waste principle,", 25–26
Biochemical oxygen demand (BOD), 150, 155–156
Biocomposites, feather keratin as, 103
Biodegradable packaging systems, 248
Biodiesel, 111
  feather keratin in production, 103
Biogas, 108, 163–164
Biological degradation, 160–161
Biological treatment, feather waste, 98–100
Biomaterial development, feather keratin in, 100–101
Biomedical applications, feather keratin in, 101–102
Biotanin, 157
Blood, 113
BOD. *See* Biochemical oxygen demand (BOD)
Bovine, 149–150

Braskem, 222, 236*t*
Bridgestone's industrial plant, 235–236, 236*t*
BRK, 234

C

Cabot's facility, 236*t*, 237
  water balance, 238*f*, 240*f*
  water use index, evolution of, 241, 241*t*
Calcium oxide, 111–112
Camacari's Petrochemical Complex, 236
Camacari unit, 222–223
Capivari II, 232–233
Carbon black, 237
Catalysts, 111–112
CE. *See* Circular economy (CE)
Cellular-binding motifs, 101–102
Certification company, 31
Chemically reuse of manufacturing waste, 52
Chemical oxygen demand (COD), 150, 155–156
Chemicals, recovery of, 86
Chemical treatment, feather waste, 98
Chicken, 95–97
Chicken feather hydrolysis, 98
Chicken meat production, 91–93
Chrome tanned shavings, 164–165
Chrome tanning, 146–147, 161–162
Chromium, 160–161
Circular economy (CE), 7, 124, 126–127
Circular integrated waste management systems (CIWMS), 189
Circular zero-waste economy, 190, 199
CIWMS. *See* Circular integrated waste management systems (CIWMS)
Cleaner production, 153–158
  post-tanning process, 158
  pre-tanning process, 154–156
    enzymatic degreasing process, 155
    enzymatic dehairing of skin/hide process, 155–156
    enzymatic liming process, 155
    enzymatic soaking process, 154–155
  tanning process, 156–158
    enzyme-assisted three-step tanning, 156–157
    high exhaust chrome tanning, 157–158
Coca-Cola, ZW development in, 78
COD. *See* Chemical oxygen demand (COD)
Collaborative consumption, 80
Collaborative ownership, 80
Collagen fibers, 146–147
Collagen hydrolysates, 162–164
Commercial poultry farms, 91–93
Commercial waste, 45–46
Consumption of resources, 81
Conventional sludge treatment plant, 225, 226*f*
Counter-current soaking, 166, 167*f*
Cradle-to-cradle approach, 72–73
Cradle-to-cradle—zero-waste economic model, 199
Cradle-to-grave approach, 72
Crop production, 91–93
Cryogenic-assisted machining, 58–59
Culled birds, 104–105
Cysteine, 95–97

D

Dead embryos, 104–105
Degradable biocomposites, 103
Delimed fleshings, 162
Deliming and bating, 168
Dermatophytes, 99
Domestic wastewater, 238
Dried poultry waste, 108
Dry machining, 58–59
DuPont, ZW development in, 78
Dyeing process by using polymer, 158

E

Eco-city, 183
Eco-efficiency, 25–26
Eggs, 110–111, 114
Eggshell, 110–111, 110*f*
  membrane, 112
  waste, 110–112
Electric arc furnaces (EAFs), 54
Electricity based recovery, 87
Electronic gadgets, 84
Electronics, 84
Energy
  from leather industry, 163–164
  synthesis, 136
Ensiling, 104–105

Environmental disasters, 84
Environmental hazards, 247–248
Environmental Protection Law (1963), 7–8
Environmental remediation, feather keratin in, 102
Enzymatic degreasing process, 155
Enzymatic dehairing of skin/hide process, 155–156
Enzymatic liming process, 155
Enzymatic soaking process, 154–155
Enzymatic treatment, feather waste, 98–99
Enzyme-assisted degreasing technology, 155
Enzyme-assisted three-step tanning, 156–157
Enzymes, 154–155
E-waste, 2–3
  recovery, 84–85
Extracted keratin, 101–102

F

Farmyard manure (FYM), 107
Fashion brands, 248
Fashion industry, 248
Feather fibers, 103
Feather hydrolysis, 98
Feather keratin, 95–97, 97*f*, 100–101
  applications
    in agriculture, 101
    as biocomposites, 103
    in biodiesel production, 103
    in biomaterial development, 100–101
    in biomedical applications, 101–102
    in environmental remediation, 102
    in feedstock, 102
    as flame retardants, 103
    in leather processing, 103
    in textile, 102
    in wood adhesives, 100
Feather waste, 95–103
  biological treatment, 98–100
  chemical treatment, 98
  feather keratin (*see* Feather keratin)
  physical treatment, 97
Feedstock, feather keratin in, 102
Fibronectin, 101–102
Filth and Cleaning Law (1973), 7–8
Finishing operations, leather production, 147
"5R" waste hierarchy, 188–189
Flame retardants, feather keratin as, 103
Fleshings, 145–146, 162
Fleshing scraps, 163–164
Flocculating agents, 238–239
Fly ash, 70–71
Food waste, recycling of, 56–57
Fresh water index, 243
FYM. *See* Farmyard manure (FYM)

G

GBCI's true zero waste, 35–36
Gelatins, 164
Generalized remanufacturing, 52–53
Generation of waste, 1
Glendale's water reclamation plant, 220
Global Reporting Initiative (GRI), 126–127
Global waste issues, 187–192
Global waste management, 81
Global Waste Management Outlook (GWMO), 189
Glutamine, 98
Governance, zero waste, 186
Graphene oxide, 101–102
Green Circle, 37
Green urbanism, 200–201
GRI. *See* Global Reporting Initiative (GRI)
Guandu Water Treatment Plant, 233

H

Hatchery waste, 103–105, 104*f*
Hazardous wastes, 2
Hide fleshings, 162
High-density polythene (HDPE), 55–56
High exhaust chrome tanning, 157–158
Holistic model of zero waste, 73–75, 74*f*
Hollow fibers, 102
Human civilization, 2
Hydrolyzed feathers, 101
Hydrolyzed keratin, 100–101
Hydrolyzed protein, 161–162
Hydrometallurgy, 86
Hyperion Water Reclamation Plant, 220

I

ICMS, 233–234
Illegal dumping, 83
Improper disposal of poultry wastes, 94–95
Incineration, 3, 29, 159
In-convenient materials, 133–134
India, ZW development in, 78
Industrialization, 70
Industrial waste, 51, 147
  generation in 2011, 82–83, 83*t*
  producers, 84
  recovery of, 82–83
Industrial wastewater, 238
Industry 4.0, 14–15, 15*f*, 247–248
Informal electronic waste recycling, 85
Innovative technologies, 13–17
Inorganic fertilizers, 107
Integrated waste management systems (IWMSs), 184, 201
Internal reuse project, 240*f*, 241, 243
International Council of Chemical Associations (ICCA), 222
International Solid Waste Association (ISWA), 189
IWMSs. *See* Integrated waste management systems (IWMSs)

K

Keratinase enzymes, 99–100
Keratin-based composites, 101–102
Keratin film, 101–102
Keratin hydrolysate, 103, 162–163

L

Landfilling, 2–3, 25–26, 136
LCA. *See* Life cycle assessment (LCA)
Leather
  boards, 164–165
  production processing steps, 145–147
    finishing, 147
    post-tanning, 147
    pre-tanning, 145–146
    tanning, 146–147
  solid waste management, 162–163
Leather industry, 144–145
  technological solutions for, 153–169
    cleaner production (*see* Cleaner production)
    liquid waste treatment and water management (*see* Liquid waste)
    solid waste treatment and management (*see* Solid waste)
Leather industry waste, 147–152
  solid wastes, 149–150
  volatile organic compounds, 151–152
  wastewater, 150–151
Leather-processing, 147
  environmental impact of, 147, 149*f*
  feather keratin in, 103
  waste generated during, 147, 148*f*
Life cycle assessment (LCA), 17, 126
  of urban waste management, 135–136
Liming, 166–168
Lipase enzymes, 155
Liquid waste, 103–104
  treatment, 165–166
  water management approaches, 166–169
    deliming and bating, 168
    liming, 166–168
    pickling and tanning, 168–169
    soaking, 166
    washings, 169
Livestock, 91–93

M

Machining, 58
Manufacturing waste, 45–46, 48, 48*f*
MDM. *See* Mechanically deboned meat (MDM)
Meat production, 91–93
Mechanically deboned meat (MDM), 113
Mechanical treatment, feather waste, 97
Membranes, 112–113
Microbes, 98–99
Microbial treatment, feather waste, 98–99
Microwave-assisted pyrolysis, 57, 58*f*
MIFA. *See* Municipal incinerator fly ash (MIFA)
Minerals from agriculturally produced waste, 56
Minimal quantity lubrication (MQL), 59–60
MSW. *See* Municipal solid waste (MSW)
Municipal incinerator fly ash (MIFA), 54, 55*f*
Municipal solid waste (MSW), 1, 45–46

Municipal wastewater
physical and biological treatments, 242
reuse of, 217–220
Murray's challenge, 197–198

N

Nanofluid-based minimal quantity lubrication (NFMQL), 59–60
Nanoparticle dispersion (NPD), 157–158
Nanoparticle polymer (NPP), 158
Nanoparticles, 101–102
Nanotechnology, 59–60
National Institute of Standards and Technology (NIST) indicator, 13–14, 14*f*
Near dry machining, 58–59, 59*f*
Neologism, 191
Neredaâ technology, 234
New Zealand zero waste story, 202
Nitrogenous waste, 109
Nodule, 238–239
Nonedible by-products, 94–95
Nonedible poultry wastes, 94–95, 95*f*
Nonprotein nitrogen (NPN), 108
Nonrecyclable waste, 247–248
NPD. *See* Nanoparticle dispersion (NPD)
NPN. *See* Nonprotein nitrogen (NPN)
NPP. *See* Nanoparticle polymer (NPP)

O

Offal waste, 113
Organic chelates, 160
Organic matter of poultry litter, 109–110
Ovine hair, 149–150
Oxidizing agents, 98
Oxiteno plant, 236–237, 236*t*

P

Paranapanema, 235–236, 236*t*
Particulate emissions, 151
Peroxochromates, 160
Physical treatment, feather waste, 97
Pickling and tanning, 168–169
Piggery, 109–110
Plastic-based packaging materials, 248
Plastic debris, 2
Plastics, 60
Plastic waste into liquid fuel, 57–58
PM. *See* Poultry manure (PM)
P-N based flame retardant, 103
Pollution prevention (P2) assessment, 71
Polyvinyl alcohol, 102
Post-tanning process, 147
Poultry farm, 91–93, 92*f*
Poultry industry, 93
Poultry litter, 105–110
Poultry manure (PM), 105–110, 106*f*, 108*t*
Poultry meat, 114
Poultry processing steps and wastes generated, 93–95, 94*f*
Poultry production, 91–93
Poultry waste, 93
improper disposal of, 94–95
nonedible, 94–95, 95*f*
utilized for production of value-added products, 95, 96*t*
Pre-tanning process, 145–146, 150, 154–156
enzymatic degreasing process, 155
enzymatic dehairing of skin/hide process, 155–156
enzymatic liming process, 155
enzymatic soaking process, 154–155
Programmable logic controllers (PLCs), 228–229
Protein, 160–162
hydrolysates, 164
molecules, 100

R

Rainwater, 239–240
Recovery
of chemicals, 86
of industrial waste, 82–83
zero waste manufacturing, 52
Recycled water
Aquapolo's, 230–231
chemical industry and, 242
taxation of, 233–234
types, 233
Recycling, 12–13, 25–26
of food waste, 56–57
of manufacturing waste, 50
of metals, 85–86
of resources, 133–134
Redesigning, zero waste manufacturing, 48–49

Remanufacturing, zero waste manufacturing, 52–53
Residual disposal, 189
Reuse, 25–26, 124–127
  of manufacturing waste, 51

S

Sabesp, 234
San Francisco, ZW development in, 77
Seminar of Sustainable Practices, 235
Shafdan system, 219–220
Silica deposits, 229–230
Slaughter wastes, 113
Sludge treatment plant, conventional, 225, 226*f*
Soaking, 166
Solid waste, 24–25, 103–104, 149–150
  management of recovered material
    adhesives/cosmetics/films, 164
    animal feed/chicken feed/fertilizer, 161–162
    biological uses, 164
    for energy, 163–164
    in leather boards, 164–165
    in leather processing, 162–163
  theory, 199
  treatment, 159–161
    alkaline hydrolysis, 160
    biological degradation, 160–161
    incineration, 159
Solid Waste Association of North America (SWANA), 195
Solid waste management (SWM), 70, 72, 196–197
Spain, ZW development in, 77
Spent hens, 113
SS. *See* Suspended solids (SS)
Stakeholders in 3R model, 76
Stormwater, 239–240
Subaru, ZW development in, 78
Sulfur oxides, 57
Suspended solids (SS), 150
Sustainability, 11–13, 129–130
Sustainable development, 11–12, 152

T

Taiwan, zero waste in, 138
Take-make-waste, 190
Tanned hides/skin, 146–147
Tanneries, 151
Tanning agents, 146–147
Tanning process, 146–147, 150, 156–158
  enzyme-assisted three-step tanning, 156–157
  high exhaust chrome tanning, 157–158
Tannins, 161
TDS. *See* Total dissolved solids (TDS)
Technological innovations, 13–14
Tertiary Membrane Bioreactor (TMBR), 225–228, 228*f*
Textile, feather keratin in, 102
Thermal treatment, feather waste, 97
Thermoplastic films, 102
Thermoplastics, 133–134
Three-dimensional printing for waste recycling, 55–56
3R model, 72, 79–80
  in Bangladesh, 75–77
    partnership building between stakeholders, 76–77
    pilot study, 76
    role of stakeholders, 76
Tilapia, 109
Total dissolved solids (TDS), 150, 155–156, 239
Total quality management (TQM), 79
Total suspended solids (TSS), 239
Total water index, 241–242
"TRUE" ZWC program, 28
Two-stage anaerobic digestion, 104–105

U

UL environment certification, 36
Urban planning, 69–70
Urban waste management, life cycle assessment of, 135–136
Uric acid, 108

V

Vegetable tanning, 146–147
Volatile organic compounds, 151–152

W

Washings, 169
Waste, 1–4, 24–25, 70, 185
  ceramics, 60
  concept of, 127–128
  for construction, 60–61

feathers, 95–97
flow in Adelaide, 136, 137*f*
generation, 45–46, 81
global problem, 195–196
issues, 188
during leather processing, 147, 148*f*
legislative and measures, 134
materials, 81
minimization, 12–13
prevention, 182, 247
products, 53
Waste hierarchy index (WHI), 13
Waste management (WM), 69–70, 132–133, 182
in Adelaide, Australia, 136–137
governance and infrastructure, 186
hierarchy of, 11–12, 12*f*, 127–128
market-driven, 186
model by Biogen Cambridge, 124–125, 125*f*
performance of, 203
regulatory strategies, 186
by Scottish Government, 132–133, 133*f*
source of greenhouse gas emissions, 3
systems, 24–25
Waste recycling, three-dimensional printing for, 55–56
Waste reduction, 80
Waste to energy (W2E), 29, 194–195
Waste-to-energy waste (WTE), 80–81
Wastewater, 150–151
Water management approaches, 166–169
deliming and bating, 168
liming, 166–168
pickling and tanning, 168–169
soaking, 166
washings, 169
Water purification, 165–166
Water reclamation scheme, 219–220
Water Security A List, 236
Wet blue leather waste, 161–162
WHI. *See* Waste hierarchy index (WHI)
White Martins, 236–237, 236*t*
WM. *See* Waste management (WM)
Wood adhesives, 100
Wood-based waste material, 56

Z

"Zero Brine" project, 216–217
Zero liquid discharge (ZLD), 215–217
applications, 216–217
Zero waste (ZW), 4, 247
approaches, 152–153
challenges, 197–198
in electricity based recovery, 87
environmental concern, 83–84
e-waste recovery, 84–85
in recovery of chemicals, 86
recovery of industrial waste, 82–83
in recycling of metals, 85–86
circular economy, 7, 189
circularity and critique, 194–196
concept of, 71–75, 123–127, 184
convergence, 194–196
definition of, 5, 24–27
and design of future cities, 200–203
development, 75–79
3R model in Bangladesh, 75–77
in Australia, 77
in Coca-Cola, 78
in DuPont, 78
in India, 78–79
in San Francisco, 77
in Spain, 77
in Subaru, US, 78
downstream expression, 198–199
evolution of, 6–10
factors influencing multiuse of products, 128–129, 129*f*
formation, 194–196
hierarchies, 127–135
holistic model of, 73–75, 74*f*
inter-connected key principles, 129, 130*f*
key aspects of developing, 79–87
management and treatment, 132
movement, 191
of municipal incinerator fly ash, 54
phases for development, 129–130, 131*f*
phases in implementing, 8–9, 10*f*
present age of, 10–17
innovative technologies, 13–17
sustainability, 11–13
principles, 129
refined hierarchy for, 134, 135*f*

Zero waste (ZW) *(Continued)*
steps in developing, 7–8, 9*f*
strategy, 27
sustainable behavior, 127–135
in Taiwan, 138
"3R rule,", 72
trends of Africa, 137
upstream sphere, 198–199
vision, 129
waste treatment and disposal, 185–186
ZWIA's definition, 193
Zero-waste approach in handling poultry waste, 95–113
egg shell waste, 110–112
eggshell membrane, 112
feather waste, 95–103
biological treatment, 98–100
chemical treatment, 98
feather keratin (*see* Feather keratin)
physical treatment, 97
hatchery waste, 103–105, 104*f*
offal waste, 113
poultry litter and manure, 105–110
spent hens, 113
Zero waste best practice communities, 34–35
Zero waste certification (ZWC), 27–28
benefits of, 30–31
Boulder, CO, 39
Canberra, Australia, 37–38
certification company, 31
certified companies, 32
communities in California, 39
Italy, 40
New Zealand, 38
Nova Scotia, Canada, 38–39
Prince Edward Island, Canada, 39
process for certification, 32
standard, 31–32
Toronto, Canada, 39
Zero Waste International Alliance (*see* Zero Waste International Alliance (ZWIA))
Zero-waste cities, 181–182
principles of, 184–185, 184*f*
waste management, 185
Zero waste hierarchy, 31
Zero waste index (ZWI), 203
Zero Waste International Alliance (ZWIA), 26, 33, 191
categories of recognition, 33–35
minimum criteria for zero waste group, 33
zero waste community recognition, 33
Zero waste international allowance, 33–35
Zero waste manufacturing (ZWM), 45–46
industrial implications, 54–61
municipal incinerator fly ash, 54
novel methods in machining, 58–60
plastic waste into liquid fuel, 57–58
producing minerals from agriculturally produced waste, 56
recycling of food waste, 56–57
three-dimensional printing for waste recycling, 55–56
waste for construction, 60–61
prospective of, 48–53
recovery, 52
recycling, 50
redesigning, 48–49
remanufacturing, 52–53
reuse, 51
Zero-waste products, 185
Zero waste technology (ZWT), 7
degree of demonstration, 7
designing for upstream, 7
evolution of, 6–10
history of, 4–10
lifecycle design, 7
manifestation and reusability, 7
Zero waste to landfill (ZWtL), 29, 194–195
ZW. *See* Zero waste (ZW)
ZWC. *See* Zero waste certification (ZWC)
ZWI. *See* Zero waste index (ZWI)
ZWIA. *See* Zero Waste International Alliance (ZWIA)
ZWM. *See* Zero waste manufacturing (ZWM)
ZWT. *See* Zero waste technology (ZWT)

Kolibri 360 GmbH
Ferdinand-Jühlke-Straße 7
99095 Erfurt, Deutschland
produktsicherheit@kolibri360.de